A Textbook of Molecular Ecology and Environmental Engineering

Edited by **Neil Griffin**

SYRAWOOD
PUBLISHING HOUSE

New York

Published by Syrawood Publishing House,
750 Third Avenue, 9th Floor,
New York, NY 10017, USA
www.syrawoodpublishinghouse.com

A Textbook of Molecular Ecology and Environmental Engineering
Edited by Neil Griffin

Contents

Preface   **VII**

Chapter 1   **Identification of Genes Putatively Involved in the Biosynthesis of Antitubercular Peptide in *Streptomyces ribosidificus* NRRL B-11466**
Khaled M. Aboshanab, Nisreen M. Okba, Tarek S. El-banna and Ahmed A. Abd El-Aziz   **1**

Chapter 2   ***In vitro* Propagation and GC-MS Studies of *Ocimum basilicum* Linn. var. *pilosum* (Willd.) Benth**
Gaddaguti Venu Gopal, Srideepthi Repalle, Venkateswara Rao Talluri, Srinivasa Reddy Ronda and Prasada Rao Allu   **12**

Chapter 3   **Heavy Metals Concentration in Fish *Mugil cephalus* from Machilipatnam Coast and Possible Health Risks to Fish Consumers**
P. V. Krishna, K. Madhusudhana Rao, V. Swaruparani and D. Srinivas Rao   **24**

Chapter 4   ***Listeria* spp. in Raw Cow and Goat Meat in Port Harcourt, Nigeria**
O. C Eruteya, S. A Odunfa and J. Lahor   **34**

Chapter 5   **Screening of Filamentous Fungi for Xylanases and Cellulases Not Inhibited by Xylose and Glucose**
L. F. C. Ribeiro, L. F. Ribeiro, J. A. Jorge and M. L. T. M. Polizeli   **45**

Chapter 6   **Changes in Microbial Population of Palm Oil Mill Effluent Polluted Soil Amended with Chicken Droppings and Cow Dung**
L. O. Okwute and U. J. J. Ijah   **55**

Chapter 7   **Use of Palm Oil Mill Effluent as Medium for Cultivation of *Chlorella sorokiniana***
Charles Ogugua Nwuche, Doris Chidimma Ekpo, Chijioke Nwoye Eze, Hideki Aoyagi and James Chukwuma Ogbonna   **65**

Chapter 8   **Identification of Bacterial Population of Activated Sludge Process and Their Potentials in Pharmaceutical Effluent Treatment**   77
Farrokhi Meherdad, Ghaemi Naser, Najafi Fazel, Naimi-Joubani Mohammad, armanbar Rabiollah and Roohbakhsh Joorshari Esmaeil

Chapter 9   **Different Methods for DNA Extraction from Yeast-*Candida famata* Isolated from Toddy**   85
T. Santra, S. K. Ghosh and A. Chakravarty

Chapter 10   **Activity of β-Amylase in Some Fungi Strains Isolated from Forest Soil in South-Western Nigeria**   95
O. A. Oseni and M. M. Ekperigin

Chapter 11   **Biosorption of Lead by *Pleurotus florida* and *Trichoderma viride***   104
A. S. Arun Prasad, G. Varatharaju, C. Anushri and S. Dhivyasree

Chapter 12   **Effects of Utilization of Crushed, Boiled and Fermented Roselle Seeds (*Hibiscus sabdariffa*) on the Performance of Broiler Chickens**   117
Maikano Mohammed Ari, Danlami Moses Ogah, Idris Danladi Hassan, Ibrahim Suleiman Musa-Azara, Nuhu Dalami Yusuf and Samuel Emmanuel Alu

Permissions

List of Contributors

Preface

Molecular ecology is an emerging field of study that focuses on crucial challenges of ecological and environmental conservation like assessment and protection of biodiversity and species, analysing behavioural ecology, etc. It involves the use of genetics and genomics for evaluating and addressing these problems. Some of the topics covered in this extensive book are cell biology, genetics, microbial population, microbial and environmental biotechnology, applications of bioremediation and biodegradation, etc. The aim of this book is to serve as a resource guide for students and experts alike.

The researches compiled throughout the book are authentic and of high quality, combining several disciplines and from very diverse regions from around the world. Drawing on the contributions of many researchers from diverse countries, the book's objective is to provide the readers with the latest achievements in the area of research. This book will surely be a source of knowledge to all interested and researching the field.

In the end, I would like to express my deep sense of gratitude to all the authors for meeting the set deadlines in completing and submitting their research chapters. I would also like to thank the publisher for the support offered to us throughout the course of the book. Finally, I extend my sincere thanks to my family for being a constant source of inspiration and encouragement.

Editor

Identification of Genes Putatively Involved in the Biosynthesis of Antitubercular Peptide in *Streptomyces ribosidificus* NRRL B-11466

Khaled M. Aboshanab[1*], Nisreen M. Okba[2], Tarek S. El-banna[3] and Ahmed A. Abd El-Aziz[3]

[1]Department of Microbiology and Immunology, Faculty of Pharmacy, Ain Shams University, Organization of African Unity St., POB: 11566, Abbassia, Cairo, Egypt.
[2]Department of Microbiology and Immunology, Faculty of Pharmacy, Al-Azhar University (Girls), Cairo, Egypt.
[3]Department of Pharmaceutical Microbiology, Faculty of Pharmacy, Tanta University, Tanta, Egypt.

Authors' contributions

Author KMA designed the study, performed the sequence analysis, wrote the protocol, and first draft of the manuscript. Author NMO performed the practical experiments of phenotypic detection. Authors TSE and AAA managed the literature searches. All authors read and approved the final manuscript.

ABSTRACT

Aims: To determine the potential antitubercular activity of *Streptomyces ribosidificus* NRRL B-11466 both on genotypic and phenotypic levels.
Methodology: Standard methods and software programs were used for nucleotide/protein sequence analysis and phenotypic detection of antitubercular activity.
Results: Analysis of the submitted DNA segment (accession code = AJ744850) harbouring the ribostamycin biosynthetic gene cluster showed that the respective gene cluster was flanked in the upstream region by three open reading frames (ORFs), encoding putative type II thioesterase (SribL03.14c) and two nonribosomal peptide synthases (SribL03.14c and SribL03.14c). These ORFs were of high amino acid similarities (about 80%) to those located in the viomycin and related antibiotic biosynthetic gene clusters. A DNA segment harbouring three ORFs, putatively involved in

---

Corresponding author: Email: aboshanab2003@yahoo.com

capreomycidine biosynthesis was submitted into the GenBank database under the accession code HQ327309. Comparative analysis of the respective DNA segment with viomycin and related antibiotic biosynthetic gene clusters showed: firstly, location of the respective DNA segment in the neighbourhood and upstream to the ribostamycine biosynthetic gene cluster; secondly, conservation of six ORFs: SriC (putative L-arginine hydroxylase); SriD (putative L-capreomycidine synthase), SriE (putative permease) located on our submitted DNA fragment; and SribL03.14c, SribL03.15c, SribL03.16c located on the DNA fragment harboring ribostamycin biosynthetic gene cluster, among the tested biosynthetic gene clusters. Phenotypically, *S. ribosodificus* inhibited growth of *Mycobacterium smegmatis* ATCC 19420 and *Mycobacterium phlei* ATCC 11758.

Conclusion: *Streptomyces ribosidificus* NRRL B-11466 produces antimycobacterial agents and this was confirmed genotypically via detection of 6 ORFs with high amino acid similarities (about 80%) to those located in the viomycin and related antibiotic biosynthetic gene clusters as well as phenotypically by determining its inhibitory activity against *Mycobacterium smegmatis* ATCC 19420. This is the first report about identification of genes putatively involved capreomycidine biosynthesis in *Streptomyces ribosidificus* NRRL B-11466.

Keywords: Tuberactinomycins; capreomycidine biosynthesis; viomycin; Streptomyces ribosidificus NRRL B-11466.

1. INTRODUCTION

Tuberactinomycin family of peptide antibiotics, (tuberactinomycins; TUBs) is active against *Mycobacterium tuberculosis* infections and is particularly used for the treatment of multidrug-resistant tuberculosis, methicillin-resistant *Staphylococcus aureus* (MRSA) strains and vancomycin-resistant enterococci (VRE) [1,2]. They are peptide antibiotics characterized by the presence of capreomycidine, a nonproteinogenic amino acid with a 6-membered cyclic guanidine side chain that is biosynthesized and condensed with other amino acids via a nonribosomal peptide synthase mechanism to form various TUBs [3]. TUBs include antibiotics such as viomycin, tuberctinomycins, streptothricin and capreomycins that are produced by different *Streptomyces* strains [4-8]. The antibiotic viomycin (tuberactinomycin B), the well-studied antibiotic contain unusual amino acids such as L-capreomycidine, 2,3-diaminopropionate, β-ureidodehydroalanine, and β-lysine [3,8]. The complete biosynthetic pathway of these antibiotics still not biochemically identified, however it was anticipated that they are synthesized via a nonribosomal peptide synthase (NRPS) mechanism [8,9]. The full biosynthetic gene cluster of viomycin antibiotic from *Streptomyces* strain ATCC 11861 was completely isolated and analyzed [3,8]. The unusual nonproteinogenic amino acids were anticipated to be synthesized from α- amino acids in the cell such as 2,3-diaminopropionate from L-serine and L-ornithine. 2,3-diaminopropionate would be further modified to form, β-ureidodehydroalanine, L-capreomycidine from L-arginine, and β-lysine from L-lysine [8]. These amino acids would be condensed to produce these antibiotics via nonribosomal peptide synthases (NRPSs) whose respective genes were also located with the identified biosynthetic gene clusters.

Moreover, conversion of (2S)-arginine to (2S,3R)-capreomycidine by VioC and VioD from the viomycin biosynthetic pathway of *Streptomyces sp.* strain ATCC11861 was biochemically analyzed [10]. TUBs also target the catalytic RNAs involved in viral replication [11,12]. Interestingly, some members of TUBs family are listed in the World Health

Organization's model drug list 2002. Recently, it was investigated that tuberactinomycins inhibit translocation on 70S ribosome by stabilizing the tRNA in the A site in the pretranslocation state [13] which is adjacent to the binding sites for the some 2-deoxystreptamine aminocyclitol aminoglycoside antibiotics (2DOS-ACAGA) such as paromomycin and hygromycin B [13]. *Streptomyces ribosidificus* NRRL B-11466 is a producer of ribostamycin, a 2DOS-ACAGA. The ribostamycin biosynthetic gene cluster was completely sequenced and analysed [14]. Analysis of the submitted DNA segment harbouring the ribostamycin biosynthetic gene cluster showed the presence of three ORFs with a very good amino acid identities (about 80%) to those located in the viomycin biosynthetic gene cluster of *S. vinaceus*. These three ORFs were putative type II thioesterase and two NRPSs, however their exact biosynthetic roles in *S. ribosidificus* were not yet known. Whether a full viomycin-related biosynthetic gene cluster is located in *S. ribosidificus* has to be explored. Moreover, isolation, sequencing and annotation of three genes putatively involved in capreomycidine biosynthesis from *Streptomyces ribosidificus* NRRL B-11466 were carried out and submitted into the GenBank database under accession code HQ327309 [15]. Therefore, in this work, comparative analysis of ORFs located on the DNA segment (HQ327309) with the viomycin and capreomycin biosynthetic gene clusters was carried out. Also, preliminary antitubercle inhibitory activity of *S. ribosodificus* was tested against *Mycobacterium smegmatis* ATCC 19420 and *Mycobacterium phlei* ATCC 11758 standard strains.

2. MATERIAL AND METHODS

2.1 Bacterial strains, culture media

Streptomyces ribosidificus NRRL B-11466 (ribostamycin producer) was cultured on tryptic soy broth (TSB) [16,17] or on M65 composed of glucose 4.0 g, yeast extract 4.0 g, malt extract 10.0 g, agar 12.0 g, distilled water ad. 1000.0 ml, pH adjusted to 7.2 (DSMZ, Braunschweig, Germany) at 28°C. *Mycobacterium phlei* ATCC 11758 and *Mycobacterium smegmatis* ATCC 19420 were cultured onto nutrient agar and incubated for 48 hrs at 28°C.

2.2 Testing the Preliminary Antitubercular Inhibitory Activity of *S. ribosidificus*

S. ribosidificus NRRL B-11466 was inoculated into 25 ml TSB and incubated at 28°C for 48 hrs at 160 rpm. About 1 ml from the obtained growth was used for surface inoculation of either tryptic soy agar plate (TSB) or M65 agar plate. The surface inoculated plates were incubated at 28°C for 5 days. From Each plate, agar plug was obtained using a sterile cork borer and added on a surface of inoculated nutrient agar plates (10^5 CFU/ml) with standard testing strains (*Mycobaterium phlei* ATCC 11758 and *Mycobacterium smegmatis* ATCC 19420, a local clinical isolate of *Staphylococcus aureus*). The plates were incubated at 28°C for 24 hrs and the resulted inhibition zones were measured in mm.

2.3 Nucleotide Accession Code

The nucleotide sequence reported in this study was submitted in the NCBI GenBank database under the accession code HQ327309. The DNA fragment submitted to the NCBI GenBank harboured three ORFs namely SriC (putative L-arginine hydroxylase); SriD (putative L-capreomycidine synthase), SriE (putative permease), This DNA fragment was obtained via DNA sequencing of various PCR products obtained using various heterologous and homologous primers and chromosomal DNA of *S. ribosidificus* as a template. The obtained DNA sequence files were assembled into one final contig which was submitted into the NCBI GenBank under accession code HQ327309 [15]

2.4 Computer-assisted Analysis of DNA sequences

The programs used for computer-assisted analysis of nucleotide and protein sequences were Staden Package [18], FramePlot [19], Online analysis tools (http://molbiol-tools.ca/), ClustalW2 [20]. Structure of proteins and conserved domain analysis were conducted using Basic   Local   Alignment   Search   Tool   (NCBI) http://www.ncbi.nlm.nih.gov/Structure/cdd/cddsrv.cgi.

3. RESULTS

3.1 Comparative Analysis of the Submitted DNA Segment (HQ327309) with Various Viomycin-capreomycin Biosynthetic Gene Clusters

As shown in Fig. 1, a total of six ORFs (SriC, SriD, SriE located on our submitted DNA fragment; NCBI GenBank accession code = HQ327309; and SribL03.16, SribL03.15, SribL03.14 located on the DNA fragment harboring ribostamycin biosynthetic gene cluster, NCBI GenBank accession code = AJ744850) were highly conserved among the viomycin and capreomycin biosynthetic gene clusters. It was also obvious that the DNA segment (HQ327309) harboring the putative capreomycidine biosynthetic ORFs (SriC, putative L-arginine hydroxylase and SriD, putative L-capreomycidine synthase) was highly conserved (80% similarities). However, there was a gap between the respective DNA segment (HQ327309) and the ribostamycin biosynthetic gene cluster (accession code = AJ744850).

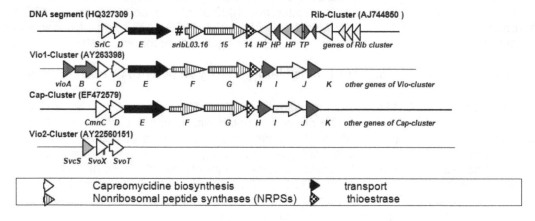

Fig. 1. Comparative analysis of the submitted DNA segment (HQ327309) with various viomycin-biosynthetic gene clusters.
Rib-Cluster = ribostamycin biosynthetic gene cluster, accession code, AJ44850; DNA segment (HQ327309)= DNA segment harbouring putative capreomycidine biosynthetic genes; Vio1-Cluster (AY263398) = Viomycin biosynthetic gene cluster, accession code AY263398; Cap-cluster (EF472579) = capreomycin biosynthetic gene cluster, accession code EF472579; Vio2-Cluster (AY22560151) = Viomycin biosynthetic gene cluster, accession code AY22560151.

3.2 Alignment of Sric (putative L-arginine hydroxylase) and Homologous Proteins

As shown in Fig. 2, SriC showed more than 80% similarities in the amino acid sequences with VioC (L-arginine hydroxylase; accession code, AAO66427) from *S. vinaceus* (viomycin producer;) and CmnC (L-arginine hydroxylase; accession code, ABR67746) from *Saccharothrix* sunsp. *capreolus* (capromycin producer). The catalytic sites were also conserved.

```
2WBQ_A      27 .[48].MARARLDAWPHALVVRGNPVDDAALGS.[3].HWRTARTPGSR     PLSFLLMLYAGLLGDVFGWATQQDGR 141
SriC         1 .[15].LRSYNLNDQTRSLKINRNDCGLRRAGP     DWRDARTPGSR     PLSFLLTLYAGLLGDVFGWATQQDGR 79
gi 6729662  12 .[48].LREFKLTDHEGHAVIRGHEFDQQRIGP.[3].DWRGRQRPGPE.[1].PEELLLMLYAALLGEPFGWATQQDGH 12
gi 150249468 7 .[48].VERARLDDRLHALVVRGNDVDQDALGP.[3].HWRQARTAASR     RYGFLLVLYASLLGDVVGWATQQDGR 121
gi 256392639 4 .[49].LDDFRLREPSALCVISGLDVDQDRLGP.[3].HWRDSQIGSRS.[1].NLEIFFLLCGAALGDVFGWATQQDGR 120


2WBQ_A     142 VVTDVLPIKGGEHTLVSSSSRQELGWHTEDAFSPYRADYVGLLSLRNPDGVATTLAGVPLDDLDERTLDVLFQERFLIRP 221
SriC        80 VVTDVLPIKGGEHTLVSSSSRQELGWHTEDAFSPHRADYVGLLSLRNPDRVATTLAGAPLDDLDERTLDVLFQDRFLIRP 159
gi 6729662 128 LVHDIFPIRQHENDQLGMGSKELLTWHTEDAFHPYRSDYLILGALRNPDRVPTTLGGLDVASLSAEDIDILFEPRFSIAP 207
gi 150249468 122VVTDVLPIEGQEDSLVSSSSSVELGWHTEDAFSPYRADYVGLFSLRNPDSVATTVAGLDPDLVGPAVVDVLFGERFHIRP 201
gi 256392639 121IMHDVLPIKGHEHYELGSNSLQHLSWHTEDSFHPCRGDYVALMCLKNPYEAETMVCDAGDLDWPNLDVDALFEPVFTQMP 200


2WBQ_A     222 DDSHLQVNNS.[5].RVE     FEGIAQAADRPEFVAILTGHRAAPHLRVDGDFSAPAEGDEEAAAALGTLRKLIDASL 296
SriC       160 DDSHLPVNNS.[3].RAR.[2].FDEIAQAVDRPEPVAVLTGHRAAPHLSVKGDFSAPAEGDEEAAAALETLRKLIEASL 234
gi 6729662 208 DESHLPKNNT.[4].EEE.[2].FATIQRMIDERPLGPLLYGSRLDPYMRLDPYFTSVPEGDTDARRAYDALYKLVDAGM 283
gi 150249468 202 DNSHLPTHNS.[2].RLS.[2].FAGIVEAVENPRAVSILRGHRDAPQLCVDSDFTTAVDGDAEAAGALDTLIKHLGGAL 275
gi 256392639 201 DNSHLPQNTA.[5].PTK.[7].FELIKSWNENPVRRAVLYGDRQNPYMALDPYHMKMDDWSERSLEAFQALCEEIEAKM 282


2WBQ_A     297 YELVLDQGDVAFIDNRRAVHGRRAFQPRYDGRDRWLKRINITRDLHRSR.[1].AW     AGDSRVL 355
SriC       235 YELVLDAGDVAFIDNRRAVHGRRAFRPRYDGRDRWLKRINITRDLHRSR     EI.[2].SGDSRVL 294
gi 6729662 284 REVVADQGDVLFIDNHRAVHGRLPFKAHYDGTDRWLKRVCVTADLRRSR     EM.[2].TAATRLL 343
gi 150249468 276 YEVVLGPGDVAFLDNRNVVHGRRPFRARFDGTDRWLKRINVTADLRKSR     AA.[2].DAQARVL 335
gi 256392639 283 QDVVLHPGDIAFIDNFRAVHGRRSFRARYDGSDRWLKRLNITRNLRGSR     AW.[2].APDDRVI 342
```

Fig. 2. Multiple amino acid sequence Alignment of L-arginine hydroxylase of *Streptomyces ribosidificus* NRRL B-11466 (SriC; ADR02786) and its homologous. *The numbers indicate the position within the corresponding proteins: 2WBQ_A = Chain A, crystal structure of VioC in complex with (2s,3s)-hydroxyarginine, accession code (AC) = 2WBQ_A; gi 6729662 = putative oxygenase of Streptomyces rochei, AC= CAB67713; gi 150249468 = CmnC of Saccharothrix mutabilis subsp. Capreolus, AC= ABR67746; gi 256392639 = hypothetical protein Caci_3456 of Catenulispora acidiphila DSM 44928, AC= YP_003114203.*

3.3 Alignment of SriD (putative L-carpreomycidine synthase) and Homologous Proteins

As shown in Fig. 3. SriD showed about 80% similarities in the amino acid sequences with VioD (L-carpreomycidine synthase; accession code. AAO66428) from *S. vinaceus* (viomycin producer;) and CmnD (L-arginine hydroxylase; accession code, ABR67747) from *Saccharothrix* subsp. *capreolus* (capromycin producer). Multiple amino acid sequence alignment also showed conservation of the amino acid residue lysine (K; position 231 within SriD) that would be necessary for the catalytic activities of the respective proteins via forming an internal aldimine bond (Schiff base linkage) with pyridoxal -phosphate (PLP).

```
gi 6729660      17 LEEWYRRHLAPDVHDISSSGVH PYTFAEIRDL      CRIPAEDLDKTVMDDSVSQGGAGIRQAIADRYAGGDAERVLVT  91
SriD            17 LEDWLRERYFQAKTDISSSGVHNYTFGELRAL.[2].ALLGTEELDRLMFRDGPSLGDERLRAAVAVRVRPGPGHTVMTT   93
gi 29469265     17 LEDWLRERYFQAKTDISSSGVHNYTFGELRAL.[2].ALLGTRELDQLMFRDGPSLGDERLRAAVAARVRPGPGHVVMTT   93
gi 150249469    11 LEDWLRERYFTARVDVSSSGVADHRLADLRRL      GGITVEELDAVVFRDGPSLGAERLRAALADRLRPGPDHVVMTA   85
gi 220682047     8 LEDWLRDYYFTAEIDISSSGVQSYSMAELRTF      TGIEYSDLDALVFDDGYSLGTPKVREAIARRWGDGDPGKVMTT   82
gi 256392634     8 LEAWMRSYYHTVDFDIGSSGVRDLSIEELCTL      CDLDLLSLKDMPIRDSESYGGSGLRAALADRWTGGDVRPVMVT   82


gi 6729660      92 HGSSEAIALTLSTLLRPGDRVVVQEGIYHSLGHYPVATGCEVTGLPAA.[3].DGEIDPEALEALITPRTAAVIVNFPHN  169
SriD            94 HGSSEALFLAFTALVRPGDEVVVATPAYHSLSALAVTAGAVLRPWPLR.[3].GFVPDLDDLRAVLTARTRLVVVNFPHN  171
gi 29469265     94 HGSSEALYLAFAALVRPGDEVVVATPAYHSLSGLATAAGASLRPWPLR.[3].GFAPDLDDLRAVLSDRTRLVVVNFPHN  171
gi 150249469    86 HGSSEALFLAMTALVRPGDEVVVPDPAYHSLSALARACGAVLRPWPVL      GAAPDPADLRALLTPRTRLVVVNFPHN  160
gi 220682047    83 VGSSEAIWLVLTALLRPGDEVVVVQPGYHSLVELAVGLECTTRIWRLD.[3].DWRPRLDELAELVTDRTRAIIVNFPQN  160
gi 256392634    83 HGSSEAIYLVMHLALEPGDEIVVVDPAYQQLHDIAAWRGVKVTRWPLL.[3].GFRADLPALRELARSRPKMIVVNFPHN  160


gi 6729660     170 PTGITLSPRGLDALTERTAATGAVLVWDAATAEIAHRWEVLPDPGVAAAHTISYGTFSKTFGLPGLRVGWAVAPKELLTA  249
SriD           172 PSGACVDPRTRADLLDLVAGSGATLVWDGAFTDLTYEHPPLADPSQDLDRVLSFGTLSKAYGLPGLRVGWCVVPRGLVPD  251
gi 29469265    172 PSGACVDPRGRTELLDLVANSQAVLLWDGAFTDLVHDHPPLAEPSQDLDRVLSFGTLSKAYGLPGLRVGWCVVPQDLVSE  251
gi 150249469   161 PTGVTVDAAVQAELLDVVGRSGAYLLWDNAFRDLVYDAPPLPEPTALGGRVLSTGTLSKAHGLPGLRVGWCVLPADLAPE  240
gi 220682047   161 PTGASVTEAELREIVAHAERVGAYLLWDGAFADLVHDSPALPDVSTLYDRGIGFNTFSKAFGLPGLRFGWCLGPADVLAD  240
gi 256392634   161 PTGRSVTSEEQSQIIEIAAEAGAWLVWDNAFGELTYTADPLPLPLARYDRSICFGTLSKSYGLAGLRVGWCLGPEELLAR  240


gi 6729660     250 TFPLRDRTTLFLSPLVELIAERAMRSADVLIGMRAAEARDNLAHLNDWVAEH      E.[2].VRWTPPEGGVCALPVF  320
SriD           252 LVRIRDYLTLTLSPLTERVAAVAVDHAHTDALIAPRLANARNNRERDAAVGS.[2].P.[2].VELPVPRGGVTAFPRF  324
gi 29469265    252 LVRIRDYLTLSLSPLVERVAAVAVEHADALITPRLTEARHNRRRVLEWAAAS      E.[2].IDCFVPRGGVTAFPRF  322
gi 150249469   241 LVRVRDYLTLSLSPLTELLAAVAVEHADELIAPRLAEATANRRRLLDWAAAH      G      VDCPAPGGGVTAFPRF  309
gi 220682047   241 CVRIRDYTTLHTAPLVELLALGVLEHAEAFLEPRLKQARANREIARDWAAAH      P.[2].VAMTLPAGGVAAFPRL  311
gi 256392634   241 MALLRDYIALYVSPVLEFFAEQAVRHADRIVGMQREHAAGNRQRLLDWAAAR      P.[2].VRLAPPDGGVAAFVEF  311


gi 6729660     321 .[7].AGPQAVEAFCRELLARHRTLLVPGTAFGAPHG      ARLGFGGP.[15].  382
SriD           325 TGHADVTGPCERLLSEHGVLVVPGRVFGHADR      IRIGFSCP.[15].  379
gi 29469265    323 TAHTDVTDLCERLLARHGVLVVPGRVFGQADR      MRIGFSCP.[15].  377
gi 150249469   310 PGVADVTPLCDRLMSEHGVLTVPGGCFGFPDR      MRIGFGCD.[15].  364
gi 220682047   312 LGLADTYEFCENLFQQRGVLVIPGSCFGAAQH      IRLGFGGS.[15].  366
gi 256392634   312 PQHGDVTDLCRRMAEEERVLLVPGSCFGDAYA.[2].VRLGFGGS.[15].  368
```

Fig. 3. Multiple amino acid sequence Alignment of L-capreomycidine synthase of *Streptomyces ribosidificus* NRRL B-11466 (SriD; ADR02787) and its homologous.
The numbers indicate the position within the corresponding proteins.; gi 6729660 = putative aminotransferase of Streptomyces rochei, accession code (AC)= CAB67711; gi 29469265= putative L-capreomycidine of Streptomyces vinaceus (VioD), AC= AAO66428; gi 150249469= L-capreomycidine synthase (CmnD) of Saccharothrix mutabilis subsp. capreolus, AC= ABR67747 ; gi 220682047= putative L-capreomycidine synthase of Catenulispora yoronensis, AC= ACL80152; gi 256392634 = putative L-capreomycidine synthase of Catenulispora acidiphila DSM 44928, AC = YP_003114198. # = conservation of lysine amino acid (K) required for catalytic activity.

3.4 Testing the Preliminary Antitubercular Activity of *S. ribosidificus*

Results revealed that *S. ribosidificus* NRRL B-11466 was sporulated upon incubation on M65 agar while was not sporulated on TSB agar when using similar conditions of inoculation and incubations (5 days at 28°C). As shown in figures 4,5,6, the M65 agar plug of sporulated *S .ribosidificus* showed large inhibition zones (22 mm; 25mm, 27 mm) with all of the tested strains .The TSB agar plug of the non-sporulated *S. ribosidificus* showed only very weak inhibition zone (10 mm) with the local clinical isolate of *Staphyloccocus aureus* and showed no inhibition with both of the tested *Mycobacteria*.

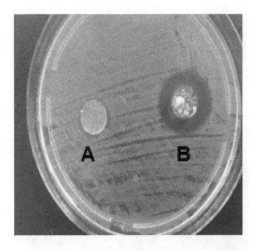

Fig. 4. Growth inhibition of *Mycobacterium smegmatis* ATCC 19420
Using :A; TSB agar plug of S .ribosidificus NRRL B-1146 (non-sporulated) B; M65 agar plug of S . ribosidificus NRRL B-11466 (sporulated)

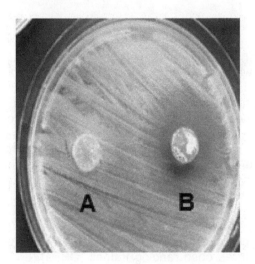

Fig. 5. Growth inhibition of *Mycobacterium phlei* ATCC 11758
Using: A; TSB agar plug of S. ribosidificus NRRL B-11466 (non-sporulated), B; M65 agar plug of S. ribosidificus NRRL B-11466 (sporulated)

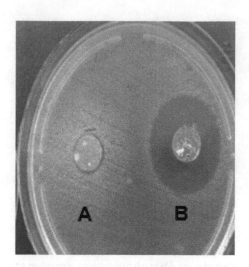

Fig. 6. Growth inhibition of *Staphylococcus aureus* clinical isolate
Using: A; TSB agar plug of S. ribosidificus NRRL B-11466 (non-sporulated), B; M65 agar plug of S. ribosidificus NRRL B-11466 (sporulated)

4. DISCUSSION

Viomycin, tuberctinomycins, streptothricin and capreomycins are major peptide antibiotics of tuberactinomycin family with enormous activity against *Mycobacterium tuberculosis* infections and are of particular importance in the treatment of the most clinically relevant pathogens such as methicilin-resistant *Staphylcoccus aureus* (MRSA) as well as vancomycin-resistant enterococci (VRE) [1,2,8,15]. The biosynthetic gene clusters of these peptide antibiotics were fully isolated and sequenced, however their complete biosynthetic pathways were not biochemically identified [3,7,8]. L-capreomycidine (amino acid with a 6-membered cyclic guanidine side chain) is the most important nonproteinogenic residue in these antibiotics was biochemically identified where VioC (L-arginine hydroxylase) and VioD (L-capreomycidine synthase) gene products were involved [10].

Analysis of the DNA fragment (NCBI accession code = AJ744850) harbouring the ribostamycin biosynthetic gene cluster showed that the respective gene cluster was flanked in the upstream region by three open reading frames (ORFs), encoding putative a type II thioestrase (SribL03.14c) and two NRPSs (SribL03.14c and SribL03.14c) [14]. These ORFs were of high amino acid similarities (about 80%) to those located in the viomycin and related antibiotic biosynthetic gene clusters [3,8,15]. In order to know, whether a full viomycin-related biosynthetic gene cluster is located in *S. ribosidificus* or not, a previous study was conducted where a series of heterologous and homologous primers were designed and used in PCR to amplify and sequence genes homologous to those in the viomycin and related antibiotic gene clusters [15]. This previous study resulted in a final assembled contig of 3884 bp which was submitted into the NCBI GenBank database under accession code HQ327309. Analysis of the respective DNA segment (contig) using FramePlot program revealed the presence of two complete ORFs (SriC, encode putative L-arginine hydroxylase and SriD, encode putative L-capreomycidine synthase) and another incomplete ORF (SriE, encode permease) [15]

Comparative analysis of the respective DNA segment with the viomycin and capreomycin biosynthetic gene clusters showed: firstly, location of the respective DNA segment in the neighbourhood and upstream to the ribostamycine biosynthetic gene cluster; secondly, conservation of six ORFs: SriC (putative L-arginine hydroxylase); SriD (putative L-capreomycidine synthase), SriE (putative permease), located on our submitted DNA fragment; and SribL03.14c , SribL03.15c, SribL03.16c located on the DNA fragment harboring ribostamycin biosynthetic gene cluster, with high amino acid identities to homologous ORFs (AAP92496.1, AAP92497.1, AAP92498.1) in the viomycin biosynthetic gene cluster [8]. This means that the presence of these genes/ORFs will be correlatated with the nature and structure of metabolic products formed by the respective clusters.

Moreover, amino acid alignment of SriC and SriD with homologous proteins together with their putative tertiary structure gave evidence about their similar catalytic activities. Thomas et al. [8] proved the essential presence of the catalytic residue lysine that forms an internal aldimine bond (Schiff-base linkage) with pyridoxal 5'-phosphate (PLP) [8,21]. This catalytic residue was also conserved in SriD (position 230). VioC and VioD proteins were biochemically analyzed to be involved in conversion of (2S)-arginine to (2S,3R)-capreomycidine [10]. Accordingly, SriC and SriD are anticipated to be involved in the biosynthesis of capreomycidine, the essential nonproteinogenic residue in the tuberactinomycin peptide antibiotics. Furthermore, conservation and arrangement of the 6 conserved ORFs by this way gave clue about presence of a peptide antibiotic biosynthetic gene cluster in a close vicinity to the ribostamycin biosynthetic gene cluster. For further confirmation, S. ribosodificus was tested phenotypically for growth inhibition of Mycobacterium smegmatis ATCC 19420 and Mycobacterium phlei ATCC 11758 standard strains as a preliminary indication of its antitubercular activity. Results showed that S. ribosidificus inhibit growth of both Mycobacterial standard strains, however the growth inhibition occurred only upon sporulation. This would means that the production of this inhibitory metabolite occurred in the stationary phase of bacterial growth which is the case of all secondary metabolites such as antibiotics. This is the first report about inhibition of Mycobacterium smegmatis growth by Streptomyces ribosidificus NRRL B-11466 as well as identification of genes putatively involved in the biosynthesis of a new peptide antibiotic of tuberactinomycin family in Streptomyces ribosidificus. Therefore, the prospective of this study is to isolate this antibiotic in a pure form, elucidate its chemical structure and confirm its activity against Mycobacterium tuberculosis in order to be used in future as antitubercular drug. Also, construction of knock-out mutant of the different genes obtained in this study followed by recording the different phenotypic changes that will occur on the mutant strain.

5. CONCLUSION

Streptomyces ribosidificus NRRL B-11466 inhibited growth Mycobacterium smegmatis ATCC 19420 and this inhibition was confirmed genotypically via isolation, sequencing and amino acid analysis of 6 ORFs with high amino acid similarities (about 80%) to those located in the viomycin and related antibiotic biosynthetic gene clusters. These ORFs were anticipated to be involved in the biosynthesis of antitubercular peptide metabolite synthesized via a nonribosomal peptide mechanism.

COMPETING INTERESTS

Authors have declared that no competing interests exist.

REFERENCES

1. Dirlam JP, Belton AM, Birsner NC, Brooks RR, Chang SP, Chandrasekaran RY, et al. Cyclic homopentapeptides 1. Analogs of tuberactinomycins and capreomycin with activity against vancomycin-resistant enterococci and Pasteurella. Bioorg Med Chem Lett. 1997;7(9):1139-44.

2. Linde Ii RG, Birsner NC, Chandrasekaran RY, Clancy J, Howe RJ, Lyssikatos JP, et al. Cyclic homopentapeptides 3. Synthetic modifications to the capreomycins and tuberactinomycins: Compounds with activity against methicillin-resistant *Staphylococcus aureus* and vancomycin-resistant enterococci. Bioorg Med Chem Lett. 1997;7(9):1149-52.

3. Yin X, O'Hare T, Gould SJ, and Zabriskie TM. Identification and cloning of genes encoding viomycin biosynthesis from *Streptomyces vinaceus* and evidence for involvement of a rare oxygenase. Gene 2003;312:215-24.

4. Carter JH, 2nd, Du Bus RH, Dyer JR, Floyd JC, Rice KC, and Shaw PD. Biosynthesis of viomycin. I. Origin of alpha, beta-diaminopropionic acid and serine. Biochemistry. 1974;13(6):1221-27.

5. Fernandez-Moreno MA, Vallin C, and Malpartida F. Streptothricin biosynthesis is catalyzed by enzymes related to nonribosomal peptide bond formation. J Bacteriol. 1997;179(22):6929-36.

6. Gould SJ, Martinkus KJ, and Tann C-H. Biosynthesis of streptothricin F. 1. Observing the interaction of primary and secondary metabolism with [1,2-13C2]acetate. J American Chemical Society. 1981;103(10):2871-72.

7. Gould SJ and Minott DA. Biosynthesis of capreomycin. 1. Incorporation of arginine. The Journal of Organic Chemistry. 1992;57(19):5214-17.

8. Thomas MG, Chan YA, and Ozanick SG. Deciphering tuberactinomycin biosynthesis: isolation, sequencing, and annotation of the viomycin biosynthetic gene cluster. Antimicrob Agents Chemother. 2003;47(9):2823-30.

9. Challis GL, Ravel J, and Townsend CA. Predictive, structure-based model of amino acid recognition by nonribosomal peptide synthetase adenylation domains. Chem Biol. 2000;7(3):211-24.

10. Ju J, Ozanick SG, Shen B, and Thomas MG. Conversion of (2S)-arginine to (2S,3R)-capreomycidine by VioC and VioD from the viomycin biosynthetic pathway of *Streptomyces* sp. strain ATCC11861. Chembiochem. 2004;5(9):1281-85.

11. Jenne A, Hartig JS, Piganeau N, Tauer A, Samarsky DA, Green MR, et al. Rapid identification and characterization of hammerhead-ribozyme inhibitors using fluorescence-based technology. Nat Biotechnol. 2001;19(1):56-61.

12. Rogers J, Chang AH, von Ahsen U, Schroeder R, and Davies J. Inhibition of the self-cleavage reaction of the human hepatitis delta virus ribozyme by antibiotics. J Mol Biol. 1996;259(5):916-25.

13. Stanley RE, Blaha G, Grodzicki RL, Strickler MD, and Steitz TA. The structures of the anti-tuberculosis antibiotics viomycin and capreomycin bound to the 70S ribosome. Nat Struct Mol Biol. 2010;17(3):289-93.

14. Piepersberg W, Aboshanab KM, Schmidt-Beißner H, and Wehmeier UF. The Biochemistry and Genetics of Aminoglycoside Producers. In: Aminoglycoside Antibiotics: John Wiley & Sons, Inc.; 2007; 15-118.

15. Aboshanab KM. Isolation, sequencing and annotation of three genes putatively involved in capreomycidine biosynthesis in *Streptomyces ribosidificus* NRRL B-11466. Eg J Med Microbio 2010;19(4):63-73.

16. Hopwood DA and Wright HM. Bacterial protoplast fusion: recombination in fused protoplasts of *Streptomyces coelicolor*. Mol Gen Genet. 1978;162(3):307-17.

17. Kieser T and John Innes Foundation., Practical *streptomyces* genetics. Norwich: John Innes Foundation. 2000.

18. Staden R. The Staden sequence analysis package. Mol Biotechnol. 1996;5(3):233-41.

19. Ishikawa J and Hotta K. FramePlot: a new implementation of the frame analysis for predicting protein-coding regions in bacterial DNA with a high G + C content. FEMS Microbiol Lett. 1999;174(2):251-53.

20. Thompson JD, Higgins DG, and Gibson TJ. CLUSTAL W: improving the sensitivity of progressive multiple sequence alignment through sequence weighting, position-specific gap penalties and weight matrix choice. Nucleic Acids Res. 1994;22(22):4673-80.

21. Barkei J, Kevany B, Felnagle E, and Thomas M. Investigations into viomycin biosynthesis by using heterologous production in *Streptomyces lividans*. Chem Bio Chem. 2009;10(2):366. doi:10.1002/cbic.200800646

In vitro Propagation and GC-MS Studies of *Ocimum basilicum* Linn. var. *pilosum* (Willd.) Benth

Gaddaguti Venu Gopal[1], Srideepthi Repalle[1], Venkateswara Rao Talluri[1], Srinivasa Reddy Ronda[2] and Prasada Rao Allu[1*]

[1]*Centre for Plant Tissue Culture and Breeding, Department of Biotechnology, K L University, Guntur- 522 502, A. P, India.*
[2]*Centre for bioprocess Technology, Department of Biotechnology, K L University, Guntur- 522 502, A.P, India.*

Authors' contributions

This work was carried out in collaboration by five authors. Author PRA suggested the study and extended overall guidance during the experimentation and made the final draft of the manuscript. Author GVG performed the experimental work, analyzed data and wrote the first draft of the manuscript. Author SRR designed GC-MS studies. Author VRT designed in vitro propagation protocol and author SR managed the literature searches and performed the statistical analysis. All authors read and approved the final manuscript.

ABSTRACT

An *in vitro* propagation method is outlined for *Ocimum basilicum Linn. var. pilosum* (Willd.) Benth., a wild aromatic plant belongs to Lamiaceae family. Shoot buds were used as source of explants on MS media supplemented with different concentrations of growth regulators for callus growth, induction of multiple shoots and roots respectively. MS media with 1.5 mg/L of kinetin and 0.5 mg/L of NAA showed 95.5% shooting, maximum number of shoots (7.33) and relatively better shoot lengths (4.15 cm). Excised shoots were carefully transferred to half-strength MS medium supplemented with 1.0 mg/L indole-3-butyric acid (IBA) for root induction and it yields 86.6% rooting. Whereas, average root length and number of roots observed were 1.73 cm and 3.31 respectively per explants. Rooted plantlets were hardened and successfully established in natural soil, where they grew and matured normally. GC-MS studies in methanolic leaf extract of naturally grown *Ocimum* species yielded 15 compounds.

*Corresponding author: Email: prallu_2006@rediffmail.com

Two compounds viz. cis-9-Hexadecenal (35.06%) and n-Hexa decanoic acid (21.6%) accounted for the major share (56.66%). On the other hand, 2-hydroxy-6-methylbenzaldehyde (10.99%) as well as 4H-1-Benzopyran-4-one and 5-Hydroxy-6, 7-Dimethoxy-2-(4-Methoxyphenyl) (7.75%) represented only 18.74%. Establishment of reliable *in vitro* propagation protocol, with phytochemical profile of hitherto unreported *Ocimum* species further widens the scope to evaluate its therapeutic properties.

Keywords: Ocimum basilicum Linn. var. pilosum (Willd.)-Benth.; Nodal segments; In vitro propagation; GC-MS analysis; 2, 4- Dichlorophenoxy acetic acid.

1. INTRODUCTION

Plants constitute the major source of raw materials for preparation of drugs. Due to wide spread toxicity and harmful side effects often caused by synthetic drugs and antibiotics, modern society increasingly prefers drugs of natural origin. Thus, about 75% of the world population still depends on medicinal plants for meeting primary health care needs [1]. The success of any healthcare programme depends on the accessibility of appropriate drugs on a sustainable basis.

The genus *Ocimum* contains about 150 species and almost all are aromatic in nature with a wide range of essential oils and many of which are extensively used in traditional medicine [2] and exhibit phytotherapeutic properties [3], antimicrobial [4], antifungal [5], as well as antioxidant and insect repellent activities [6, 7]. Various parts of *Ocimum* species are used in folk medicine for treatment of a wide range of health conditions [8]. Leaves are mainly used for the treatment of cough, bronchitis, skin disease, measles, abdominal pains and diarrhea [9]. Rats exposed to electromagnetic field of sweet basil showed decrease in immobility score and increase in forced swimming [10].

Species of *Ocimum* are usually propagated by seeds although, some exhibit poor viability and low germination. Moreover, seed-derived progenies are not true to type, due to cross-pollination. In addition, genetic diversity of these plants is threatened with extinction due to over exploitation, environment-unfriendly harvesting techniques, loss of growth habitats, uncontrolled trade and indiscriminate collection of economically important species by traditional ayurvedic practitioners. Furthermore, propagation of such plants by conventional techniques like rooting of cuttings and grafting is highly inadequate to meet the ever-growing demand. In view of this, it is important to develop suitable techniques for rapid mass propagation to meet the commercial need and also for prevention of genetic erosion. Plant tissue culture techniques offer one of the best options for conservation of rare, threatened or endangered medicinal plants.

Although reports on *in vitro* propagation studies in several *Ocimum* species are available, development of efficient micro propagation protocols for many *Ocimum* species is still in progress [11]. Rapid micro propagation protocols in young leaves, node, axillary shoot, shoot tip and inflorescence explants have been reported in *O. basilicum* [12,13], *O. sanctum* [14], *O. Kilimandscharicum* [15], *O. gratissimum* [16], *O. americanum, O. americanum* and *O.canum* [17]. In *Ocimum basilicum*, direct somatic embryogenesis from explants of leaf callus has also been reported [18].

The chemical composition and essential oil distribution in a vast number of *Ocimum* species have been studied using different extraction methods including Gas chromatograpy-Mass Spectrum [19,20], dry method and extraction pressure [21], steam and microwave distillation [22] and Super critical CO_2 extraction [23] in different plant parts. In another study, the influence of polyamines on essential oil composition is also reported in *Ocimum basilicum* [24]. Although extensive research has been conducted in several species in the genus *Ocimum*, *in vitro* propagation studies in *Ocimum basilicum* Linn. var. *pilosum* (Willd.) Benth. has not been studied due to its restricted availability. The present study made an attempt to corroborate the phytochemical composition of *Ocimum basilicum Linn*.var. *pilosum* species and also establish a reliable and efficient *in vitro* propagation protocol using nodal explants for large-scale production and germplasm conservation of this plant.

2. MATERIALS AND METHODS

2.1 Plant Materials

Healthy plants were collected from natural habitat of Kondapalli reserve forest (fly ash dumping site) which lies between 16°37' N latitude and 80°32' E longitude at a height of 168 meters above mean sea level, Vijayawada of Krishna district, Andhra Pradesh, India. Plants were successfully planted in the University botanical garden for further use. Identification of plant was done at Botanical Survey of India, Southern Regional Centre, Coimbatore and the voucher specimen is preserved in the Department of Biotechnology, KL University.

Shoot tips (nodal segments of 0.8-1.0 cm with dormant auxiliary buds of about 0.6 cm) were excised from plants and used for *in vitro* propagation studies. Selected nodal segments were immediately washed in tap water followed by washing with 10% teepol (10 min) and 1 % (w/v) Bavastin for 5 minutes. The explants were then sterilised for 20 seconds with 70% ethanol, followed by 0.1% (w/v) $HgCl_2$ for 2-3 minutes and rinsed 5-7 times in sterile distilled water in laminar air flow chamber. Thereafter, explants were trimmed at the cut ends and inoculated.

2.2 Culture Medium and Conditions

Shoot tips were cultured on modified MS basal medium [25] containing 3% (w/v) sucrose for callus initiation. The pH of the medium was adjusted to 5.7 with 1 N NaOH or 1 N HCl before gelling with 0.8% (w/v) agar. Vertically implanted explants in test tubes were maintained at $25 \pm 2°C$ temperature, continuous light with a 16 hours photoperiod at 50 µmol m^{-2} s^{-1} irradiance by cool white fluorescent tubes with 60-70% relative humidity.

2.2.1 Callus induction

Two week old sub cultured shoot buds were inoculated on modified MS medium supplemented with 2, 4 - D (0.1mg/L) and different concentrations of Kinetin (0.25–1mg/L) for optimization of callus.

2.2.2 Shoot bud initiation

Nodal segments of *O. basilicum* were inoculated on MS media supplemented with different concentrations of Indole-3-acetic acid, (IAA 0.1- 3.0 mg/L), Naphthalene acetic acid, (NAA

0.1- 3.0 mg/L and Kinetin (Kn 0.1-3.0 mg/L) either independently or in combination with other growth hormones.

2.2.3 Multiple shoot initiation

Fully matured shoots of around 3 - 4 cm in height were cut at their nodal segments and transferred to MS full strength medium (supplemented with Kn 1.0+NAA 0.5).

2.2.4 *In Vitro* rooting

For root induction, excised micro shoots with 3 - 4 fully expanded leaves from *in vitro* grown plants were transferred to half strength basal MS medium supplemented with different concentrations of IBA. The rooting results were taken 15 days after inoculation.

2.3 Acclimatization of Regenerated Plants

Fully rooted plantlets of nearly 5 to 6 cm in length were removed from the culture medium and washed under running sterile water to remove agar. The plantlets were transferred to poly trays containing sand, vermiculite and soil (1:1:2) and covered with transparent plastic bags to prevent loss of humidity. The set-up was maintained at 26 ± 1°C, 80 – 85% relative humidity and at a light intensity of 50 μmol m^{-2} s^{-2} under a 16 h photoperiod in culture room conditions and acclimatized for a period of 3 weeks. After primary hardening the plantlets were transferred to a greenhouse with simulated habitat for improved survival.

2.4 Statistical Analysis

The experiments were performed using completely randomized design and replicated three times. 12 to 15 explants per replicate were used in each treatment. Data were analyzed by one way ANOVA and the mean values for treatments were compared by using Turkey's HSD test at $p \geq 0.05$ with SPSS ver.13.0. The results are expressed as means ±SE of three experiments.

2.5 Isolation of Phytochemical Compounds

Fresh and fully expanded field grown *Ocimum* leaves from six week old plants were shade dried at room temperature, blended and made in to fine powder.100 gm of leaf powder was mixed with 1000 ml (1:10) of methanol in a Schott Duran bottle and kept air tight for 48 hours on magnetic stirrer with continuous stirring for proper mixing of powdered samples. The solutions were distilled and the extracts were used for GC - MS studies.

2.6 Gas Chromatography - Mass Spectrometry Analysis

Phytochemicals were analyzed by GC-MS (SHIMADZU QP 2010) employing the electron impact (EI) mode at an ionizing potential of 70 eV with a 30 m × 0.32 mm, film thickness and 1.8 μm capillary column (Resteck-624 MS) packed with 5% phenyl dimethyl silicone at an ion source temperature of 200°C. For further analysis, GC/MS settings were as follows: the initial column temperature was set at 45°C and held for 4 min; the temperature was raised to 50°C and then increased up to 175°C at a rate of 10^{0}C / min for 2 minutes, and then finally programmed to 240°C at a rate of 25°C / min, and kept isothermal for 2 minutes. Helium was used as carrier gas with a flow rate of 1.491 ml / min with a split ratio of 1:10. During sample analysis the column oven temperature was maintained at 280°C [19].

2.7 Identification of Compounds

The fraction composition of the sample was computed from the GC peak areas and compared with the spectra of compounds stored in the spectral database, NIST08s, WILEY8, and FAME libraries.

3. RESULTS AND DISCUSSION

3.1 Culture Response to Growth Regulators: Multiple Shooting

Ocimum basilicum Linn. var. *pilosum* (Willd.) Benth., plants were efficiently regenerated from nodal explants from field-grown young plants (Fig.1A), on MS medium supplemented by0.5-3 mg/L kinetin with 0.5- 2.5 mg/L NAA or 0.5-3 mg/L of IAA for multiple shoot induction. The multiple shoot induction response with respect to the test concentrations of growth hormones is presented in Table 1. Of the two combinations (Kinetin with IAA / NAA) tested in the present study, Kinetin with NAA was found to exhibit highest shooting rate and better shoot length per explants (Fig. 1B). MS media supplemented with Kn 0.5 + IAA 1.5 mg/L and Kn 0.5 mg/L in combination with 1.5 mg/L of NAA gave 84.4% and 86.7% shoots respectively. On the other hand, MS media supplemented with IAA (Kn1.5 + IAA 0.5 mg/L) showed 71% shooting with less number of shoots (2.24) and shorter shoot lengths (1.97cm) per explants. MS media supplemented with Kn 0.5 and 1.5 mg/L of NAA generated 5.96 shoots per explants whereas multiple shoot production on MS media supplemented with 0.5 mg/L of kinetin with 1.5 mg/L of IAA showed 3.58 shoots. Average number of shoots on MS medium supplemented with NAA was found to be significantly higher ($P < 0.5$) when compared to the MS media supplemented with IAA. On the other hand, shoot lengths appeared to be almost similar in the two combinations of growth hormones tested in the present study. Overall, MS media supplemented with 1 mg/L of kinetin with 0.5 mg/L of NAA showed highest multiple shoots (95.5%), maximum number of shoots (7.33) and even relatively longer shoots (4.81 cm) per explants (four weeks after ideal cultural conditions). This response was significantly better ($P < 0.05$) than that from other combinations tested in the present investigation. From the results, it is evident that the higher proportion of kinetin with half the concentration of NAA holds good for attaining rapid multiple shoots under *in vitro* conditions. Kinetin with different root induction growth hormones (auxin types and concentrations) greatly influences auxillary shoot regeneration from nodal explants. Medium without growth regulator (control) gave no regeneration response and explants swelled and became necrotic two weeks after inoculation. Multiple shoot production, number of shoots and length of the multiple shoots per explants significantly reduced in all combinations of growth hormones tested either above or below optimized concentrations. Growth response (reduction in number of shoots per each node) at higher or lower than optimal concentration of cytokine has also been reported in several medicinal plants [26-28]. Other cytokinins like Benzyl adenine have been reported to overcome apical dominance, release of lateral buds from dormancy and promote shoot formation in dicot plants [29]. Effect of BA on multiple shoot formation of *in vitro* propagated *O. basilicum* has also been reported [11,12]

Table 1. Effect of different concentrations of Kinetin (Kn) in combination with Indole Acetic Acid (IAA) and Naphthalene acetic acid (NAA) for multiple shoot induction from nodal explants of *Ocimum basilicum* Linn. var. *pilosum* (Willd.) Benth.

S. No	Growth Regulators (mg/L)	Shooting Percentage	Number of Shoots / explant	Shoot length / explant (cm)
0	Control (basal medium)	0.00	0.00	0.00
1	MS + Kn 0.5 + IAA 0.5	57.7	$1.20^{b} \pm 0.1815$	$1.17^{b} \pm 0.1978$
2	MS + Kn 1.0 + IAA 0.5	64.4	$1.36^{b} \pm 0.1963$	$1.63^{a} \pm 0.1975$
3	MS + Kn 1.5 + IAA 0.5	71.1	$2.24^{a} \pm 0.2424$	$1.97^{a} \pm 0.2068$
4	MS + Kn 2.0 + IAA 0.5	66.6	$1.98^{a} \pm 0.2389$	$1.81^{a} \pm 0.2037$
5	MS + Kn 3.0 + IAA 0.5	60.0	$1.60^{b} \pm 0.2118$	$1.38^{b} \pm 0.1764$
6	MS + Kn 0.5 + IAA 1.0	63.3	$2.60^{b} \pm 0.2551$	$2.26^{b} \pm 0.2191$
7	MS + Kn 0.5 + IAA 1.5	84.4	$3.58^{a} \pm 0.2431$	$3.47^{a} \pm 0.2305$
8	MS + Kn 0.5 + IAA 2.0	77.7	$3.29^{a} \pm 0.2746$	$2.72^{b} \pm 0.2124$
9	MS + Kn 0.5 + IAA 3.0	68.8	$2.71^{b} \pm 0.2836$	$1.89^{c} \pm 0.2077$
10	MS + Kn 0.5 + NAA 0.5	82.2	$3.78^{c} \pm 0.2996$	$3.11^{b} \pm 0.2435$
11	MS + Kn 1.0 + NAA 0.5	95.5	$7.33^{a} \pm 0.2947$	$4.81^{a} \pm 0.1736$
12	MS + Kn 1.5 + NAA 0.5	88.8	$5.58^{b} \pm 0.3474$	$4.34^{a} \pm 0.2456$
13	MS + Kn 2.0 + NAA 0.5	75.5	$2.76^{c} \pm 0.2546$	2.51 ± 0.2239
14	MS + Kn 3.0 + NAA 0.5	64.4	$2.22^{c} \pm 0.2713$	$1.67^{c} \pm 0.1978^{d}$
15	MS + Kn 0.5 + NAA1.0	80.0	$4.89^{a} \pm 0.3882$	$2.97^{b} \pm 0.2375$
16	MS + Kn 0.5 + NAA 1.5	86.7	$5.96^{a} \pm 0.3651$	$3.96^{a} \pm 0.2372$
17	MS + Kn 0.5 + NAA 2.0	75.6	$3.53^{b} \pm 0.3476$	$2.42^{b} \pm 0.2219$
18	MS + Kn 0.5 + NAA 3.0	66.7	$2.44^{c} \pm 0.2905$	$1.89^{c} \pm 0.2131$

Values are means ± SE. (n = 15). Means followed by different letters in the same column found significant at 5% level, Means followed by same letters do not differ significant at 5% level. (Turkey's HSD test).

3.2 Rooting

It is a well established concept that *in vitro* root induction in many plant species at slightly higher proportions of auxin to cytokine ratio in MS media yields better results. Per cent root induction, number of roots and root lengths obtained in this study are shown in Table 2. Generally, MS media supplemented with 1 mg/L of IBA were found to be remarkable among all root induction hormones tested in the present study. The best response with optimum rooting (86.6%) was observed in MS media containing 1 mg/L of IBA which also gave the highest number of roots (3.31) per explants and an average of 1.73 cm of root length(Fig. 1C). On the other hand relatively poor rooting response was observed on MS media supplemented with 0.5 to 2 mg/L of NAA and / or IAA. Other studies reported that optimum rooting in shoots of *O. basilicum* was achieved on half strength MS medium supplemented with 1.0 mg/1 NAA and thus support the theoretical concept of root induction in media supplemented with either NAA or IBA. Turkey's HSD test confirms P ≤ 0.05 significance among the three concentrations (0.5, 1 and 2 mg/L IBA) tested in the present investigation.

Table 2. Effect of different concentrations of Naphthalene Acetic Acid (NAA), Indole butyric Acid (IBA) and Indole Acetic acid (IAA) on root induction from nodal explants of *Ocimum basilicum* Linn. var. *pilosum* (Willd.) Benth

S. No	Growth Regulators (mg/L)	Rooting Percentage	Number of Roots / explant	Root length / explant (cm)
1	MS+ NAA 0.5	17.7	$0.53^c \pm 0.1786$	$0.36^b \pm 0.1174$
2	MS + NAA 1.0	37.7	$1.16^b \pm 0.2334$	$0.72^a \pm 0.1469$
3	MS + NAA 0.5	53.3	$1.73^a \pm 0.2589$	$0.97^a \pm 0.1494$
4	MS + NAA 2.0	44.4	$1.38^a \pm 0.2447$	$0.82^a \pm 0.1486$
5	MS + IBA 0.5	77.7	$2.67^b \pm 0.2441$	$1.46^{ab} \pm 0.1347$
6	MS + IBA 1.0	86.6	$3.31^a \pm 0.2221$	$1.73^a \pm 0.1267$
7	MS + IBA 1.5	73.3	$2.40^b \pm 0.2344$	$1.23^b \pm 0.1267$
8	MS + IBA 2.0	66.6	$1.87^c \pm 0.2146$	$1.10^{bc} \pm 0.1287$
9	MS + IAA 0.5	26.6	$0.87^a \pm 0.2216$	$0.49^b \pm 0.1266$
10	MS + IAA 1.0	33.3	$0.98^a \pm 0.2144$	$0.62^{ab} \pm 0.1396$
11	MS + IAA 1.5	42.2	$1.24^a \pm 0.2317$	$0.76^a \pm 0.1436$
12	MS + IAA 2.0	35.5	$1.09^a \pm 0.2309$	$0.69^{ab} \pm 0.1459$

Values are means ± SE. (n = 15). Means followed by same letters in the same column do not differ significantly at 5% level (Turkey's HSD test).

3.3 Callus Induction

Dedifferentiated nodal explants were transferred on to the callus induction medium supplemented with 0.1 mg/L of 2, 4-D and different concentrations (0.25 – 1 mg/L) of kinetin (Table 3). Growth of callus was measured in four week old cultures and expressed as dry biomass. Overall, highest callus dry biomass (64 mg) was recorded in MS medium supplemented with 0.1 mg/L of 2, 4-D and 0.5 mg/L of Kinetin (Fig. 1E). Other combinations tested in the present study were also found to be promising in callus growth. Generally, callus induction studies in many *in vitro* propagation protocols reveal that equal proportions of growth hormones in culture media play a crucial role in optimal callus induction. Contrary to this trend, in the present study it was found that at a relatively higher proportion of Cytokine to auxin, optimum callus biomass prevailed. Surprisingly, approximately equal proportions of auxin to cytokinin induce extensive rooting rather than callus formation in the present study. The altered rooting response in the highly aromatic *Ocimum* species is attributed to secondary metabolite production especially formation of tryptophan, the precursor for auxin synthesis (Data Unpublished).

Table 3. Effect of different concentrations of Kinetin on callus dry biomass from nodal explants of *Ocimum basilicum* Linn. var. *pilosum* (Willd.) Benth

S. No	Growth Medium	Concentration of growth regulators(mg/L)	Dry biomass of Callus (mg)
1	MS Basal medium	0.00*	0.00^{NA}
2	MS Basal medium	2,4-D (0.1), Kinetin (0.25)	54.3 (±0.5786)
3	MS Basal medium	2,4-D (0.1), Kinetin (0.50)	64.0 (±0.3527)
4	MS Basal medium	2,4-D (0.1), Kinetin (0.75)	42.0 (±0.6794)
5	MS Basal medium	2,4-D (0.1), Kinetin (1.00)	33.9 (±0.4991)

Without growth regulators; [NA] Callus growth not observed
Values are means of 10 replicated calli grown in three different culture bottles. ± SE

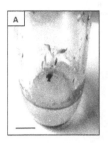

Fig. 1. *In vitro* clonal propagation of *Ocimum basilicum* Linn. var. *pilosum* (Willd.) Benth. – developmental stages (A-E).
A) Single nodal explants with induced growth of nodal buds on MS medium supplemented with 1 mg / L of kinetin (Kn) and 0.5 mg / L of Naphthalene acetic acid (NAA)-one week after inoculation. Bar = 0.5 cm. B) Shoot multiplication on MS medium supplemented with Kinetin 1.0 + NAA 0.5 mg / L 4 weeks of culture, Bar = 0.8 cm. C) Regenerated shoots with well developed adventious roots cultured on ½ MS medium supplemented with 1 mg / L of BA, Bar = 0.7cm. D) Well developed hardened plant, Bar = 4 cm.) E) Proliferation of callus from nodal segments on MS medium supplemented with 0.1 mg / L of 2,4-D with 0.5 mg/L of Kinetin, Bar = 0.5 cm.

3.4 Chemical Composition of the Essential Oils

The essential oils in shade dried leaves of *Ocimum basilicum* Linn. var. *pilosum* (Willd.) Benth. extracted in methanol were analyzed by GC-MS. The qualitative and quantitative parameters for all the 15 compounds along with per cent distribution and molecular structures are shown in Fig. 2. The distribution of phyto constituents in this species differs from what was obtained in other *Ocimum* species reported so far. Major constituents the essential oil of *O. basilicum* Linn. var. *pilosum* (Willd.) Benth. were cis-9 -Hexadecenal (35.06%) and n - Hexadecanoic acid (21.6%). Others were, 2 - hydroxyl - 6-methylbenzaldehyde (10.99%), 4H-1-Benzopyran-4-one, 5-Hydroxy-6, 7-Dimethoxy-2-(4-Methoxyphenyl) (7.75%), Phytol (4.37%), Cycloisolongifolene,7-bromo (3.31%), Neophytadiene (2.75%), Benzoic acid (2.62%), Olealdehyde (2.4%), 1,2,3 - Propanetriol and Mono acetate(2.16%). Geranic acid, p -Methylbenzoic acid, All-trans-Squalene, 1, 2 - Benzenediol and Propylure were also present in relatively low proportions. The essential oil composition and antimicrobial activity of different *Ocimum* species have been investigated in sufficiently large number of species globally [12,19,30]. Although, essential oil composition in *O. basilicum* Linn. var. *pilosum* (Willd.) Benth. has been previously reported [31], the chemical constituents in the naturally available aromatic *Ocimum* species grown in fly ash dumping location typically exhibit novel oil composition and the results contradict the earlier reports in this species. Same species grown in different climatic conditions might alter the basic mechanism for production of essential oils in order to survive in the adverse environment and thus could play a significant role in changing the chemical composition of the species. Impact of climatic conditions on altered phytochemical constituents and its adaptability to the changing environment suggests that *Ocimum basilicum* Linn. var. *pilosum* (Willd.) Benth. could be a good source for development of new and improved traits, which in turn could be exploited in the fragrance industry.

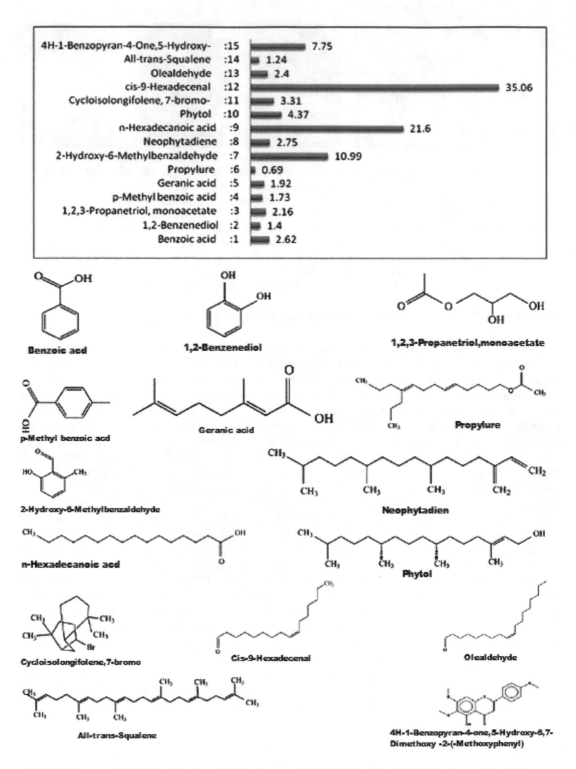

Fig. 2. Per cent distribution and structures of phytochemicals in *Ocimum basilicum* Linn. var. *pilosum* (Willd.) Benth.

4. CONCLUSION

In the present study, a competent and reliable micropropagation protocol for *in vitro* regeneration of *Ocimum basilicum* Linn. var. *pilosum* (Willd.) Benth, from nodal ex-plant has been established. This ensures large scale supply of the targeted plants through micro propagation technique, which is important for sustainability and conservation of germplasm. Further, the study also demonstrated that areal parts of *O. basilicum* Linn. var. *pilosum* (Willd.)-Benth. posses a wide range of phytochemicals capable of generating aromatic compounds of commercial importance which can meet the needs of flavoring industries.

COMPETING INTERESTS

Authors declare that there are no competing interests between individuals that can affect the publication of this work.

REFERENCES

1.   Cunningham AB. African medicinal plants, Setting priorities at the interface between conseivation and primary healthcare. People and Plants Working Paper. 1993;1-50.
2.   Javanmardi J, Khalighi A, Kashi A, Bais H & Vivanco J, Chemical characterization of basil (*Ocimum basilicum L.*) found in local accessions and used in traditional medicines in Iran, Journal of agricultural and food chemistry. 2002;50:5878-5883.
3.   Maria Magdalena Zamfirache IB, Zenovia Olteanu, Simona Dunca, Ştefania Surdu, Elena Truta, Marius Ştefan, Craita Maria Rosu, Research regarding the volatile oils composition for *Ocimum basilicum L.* and their possible phytotherapeutic effects. 2008;35-40.
4.   Ram Swaroop Verma PSB, Rajendra Chandra Padalia, Dharmendra Saikia, Amit Chauhan. Chemical composition and antibacterial activity of essential oil from two *Ocimum* sp. grown in sub-tropical India during spring-summer cropping season, Asian Journal of Traditional Medicines. 2011;6:211-217.
5.   Zhang JW, Li SK, Wu WJ. The main chemical composition and in vitro antifungal activity of the essential oils of *Ocimum basilicum Linn. var.* pilosum (Willd.) Benth, Molecules. 2009;14:273-278.
6.   Telci I, Elmastas M, Sahin A. Chemical composition and antioxidant activity of *Ocimum* minimum essential oils, Chemistry of Natural Compounds. 2009;45:568-571.
7.   Kweka EJ, Mosha F, Lowassa A, Mahande AM, Kitau J, et al. Ethnobotanical study of some of mosquito repellent plants in north-eastern Tanzania, Malaria journal. 2008;7:152.
8.   Prakash P, Gupta N. Therapeutic uses of *Ocimum sanctum Linn* (Tulsi) with a note on Eugenol and its Pharmacological actions: A short review., Indian J Physiol Pharmacol. 2005;49:125-131.
9.   Obeng-Ofori D, Reichmuth C, Bekele A, Hassanali A. Toxicity and protectant potential of camphor, a major component of essential oil of *Ocimum kilimandscharicum,* against four stored product beetles, International Journal of pest management. 1998;44:203-209.
10.   Mehdi Abdoly AF, Fatemeh Fathiazad, Arash Khaki, Amir Afshin Khaki, Arezoo Ibrahimi, Fatemeh Afshari, Hossien Rastgar. Antidepressant-like activities of *Ocimum basilicum* (sweet Basil) in the forced swimming test of rats exposed to electromagnetic field (EMF), African Journal of Pharmacy and Pharmacology. 2012;6:211-215.

11. Ahuja A, Verma M, Grewal S. Clonal propagation of *Ocimum* species by tissue culture. Indian Journal of Experimental Biology. 1982;20:455-458.
12. Sahoo Y, Pattnaik S, Chand P. In vitro clonal propagation of an aromatic medicinal herb *Ocimum basilicum* L.(sweet basil) by axillary shoot proliferation. *In vitro* Cellular & Developmental Biology-Plant. 1997;33:293-296.
13. Shahzad A, Faisal M, Ahmad N, Alatar MAA, Hend AA. An efficient system for in vitro multiplication of *Ocimum basilicum* through node culture. African Journal of Biotechnology. 2012;11:6055-6059.
14. Ramanuj R, Ramavat J, Bapodariya Y, Chariya L, Mandaliya V, et al. Rapid *in vitro* propagation of *Ocimum sanctum* L. through multiple shoot induction. Arhiv za poljoprivredne nauke. 2010;71.
15. Saha S, Dey T, Ghosh P. Micropropagation of *Ocimum Kilimandscharicum* Guerke (Labiatae). Acta Biologica Cracoviensia Series Botanica. 2010;52:50-58.
16. Gopi C, Sekhar YN, Ponmurugan P. In vitro multiplication of *Ocimum gratissimum* L. through direct regeneration. African Journal of Biotechnology. 2009;5:723-726.
17. Pattnaik S, Chand PK. *In vitro* propagation of the medicinal herbs *Ocimum americanum* L. syn. *O. canum* Sims.(hoary basil) and *Ocimum sanctum* L.(holy basil). Plant Cell Reports. 1996;15:846-850.
18. Gopi C, Ponmurugan P. Somatic embryogenesis and plant regeneration from leaf callus of *Ocimum basilicum* L. J Biotechnol. 2006;126:260-264.
19. Zhang S, Xu Y, Zhang J, Kong D, Hua M. Composition Analysis of Volatile Oil from Different *Ocimum basilicum Linn.* by GC-MS and Antimicrobial Activity. Chinese Journal of Pharmaceuticals. 2011;6:009.
20. Zhu-jun R. Analysis on Volatile Constituent of *Ocimum basilicum* Based on Gas Chromatography-Mass Spectrum. Journal of Anhui Agricultural Sciences. 2011;31:029.
21. Ren J, Wang Y, Zhou R, Deng Y, Liu X. Effect of Dry Method and Extraction Pressure on Extraction Volatile Oil in Basil. Northern Horticulture. 2011;4:013.
22. Maria Goretti de Vasconcelos Silva FJ d A M, Paulo Roberto & Oliveira Lopes F O S, c and Márcio Tavares Holanda, Composition of essential oils from three *Ocimum* species obtained by steam and microwave distillation and supercritical CO_2 extraction. ARKIVOC. 2004;6:66-71.
23. Zhou R, Wang Y, Ren J, Zhou X. Extraction of volatile oil from *Ocimum basilicum L.* by supercritical CO_2 and GC-MS analysis of the extract Journal of Hunan Agricultural University (Natural Sciences). 2010;5:026.
24. Karaman S, Kirecci OA, Ilcim A. Influence of Polyamines (Spermine, Spermidine and Putrescine) on The Essential Oil Composition of Basil (*Ocimum basilicum L.*). Journal of Essential Oil Research. 2008;20:288-292.
25. F M T a S. A revised medium for rapid growth and bioassays with tobacco tissue cultures., Physiol. Plant. 1962;15:473-497.
26. Kukreja A, Mathur A, Zaim M. Mass production of virus-free patchouli plants (*Pogostemon cablin (Blanco) Benth.*) by *in vitro* culture. Tropical Agriculture, Trinidad and Tobago. 1990;67:101-104.
27. Sen J, Sharma A. Micropropagation of *Withania somnifera* from germinating seeds and shoot tips. Plant cell, tissue and organ culture. 1991;26:71-73.
28. Vincent K, Mathew KM, Hariharan M. Micropropagation of *Kaempferia galanga* L. -a medicinal plant, Plant Cell, Tissue and Organ Culture. 1992;28:229-230.
29. George E. Plant Propagation by Tissue Culture. Part 1. The Technology. -Exegetics Limited, British Library, Edington Wilts; 1993.

30.   Kasali AA, Eshilokun AO, Adeola S, Winterhalter P, Knapp H, et al. Volatile oil composition of new chemotype of *Ocimum basilicum L.* from Nigeria. Flavour and fragrance journal. 2005;20:45-47.

31.   Ji-Wen Zhang S-K L a W-J W. The Main Chemical Composition and *in vitro* Antifungal Activity of the Essential Oils of *Ocimum basilicum Linn.* var. pilosum (Willd.) Benth. Molecules. 2009;14:273-278.

Heavy Metals Concentration in Fish *Mugil cephalus* from Machilipatnam Coast and Possible Health Risks to Fish Consumers

P. V. Krishna[1*], K. Madhusudhana Rao[1], V. Swaruparani[1] and D. Srinivas Rao[2]

[1]*Department of Zoology and Aquaculture, Acharya Nagarjuna University, Nagarjuna Nagar-522 510, Andhra Prades, India.*
[2]*Departmrnt of Biotechnology, Acharya Nagarjuna University, Nagarjuna Nagar-522 510, Andhra Prades, India.*

Authors' contributions

This work was carried out in collaboration between all authors. Author PVK designed the concentration of heavy metals study, performed the risk analysis to fish consumers, wrote the protocol, and wrote the first draft of the manuscript. Authors KMR and VS managed the analyses of the study. Author DSR managed the literature, final correct the manuscript. All authors read and approved the final manuscript.

ABSTRACT

Heavy metals are dangerous to aquatic organisms and it can be bioaccumulated in the food chain leading to diseases in humans. Cumulative effects of metals or chronic poisoning may occur as a result of long term expore even to low concentrations. The accumulation of heavy metals conditions depending upon the species, environmental conditions and inhibitory processes. Considering the human health risk due to the consumption of fish, the concentration of heavy metals (Zn, Pb, Mn, Cu, Cr and Hg) are investigated in fish samples collected from the Machilipatnam coast. The fish was examined for metal constituents are the basis on the human nutrition in the study area. These metal concentrations were exceeding the limits set by the world health organization (WHO). The study provides an insight into the potential impact of increased levels of metals in the environmental as well as estimated of the contaminated of fish tissues with metals.

*Corresponding author: Email: drpvkrishna@gmail.com

Keywords: Heavy metals; fish; human health risk.

1. INTRODUCTION

Heavy metals are stable and persist in environmental contaminants of aquatic environments and their organisms. They occur in the environment both as a result of natural processes and as pollutants from human activity [1]. According to World Health Organization (1991), metal occur in less than 1% of the earths crust, with trace amounts generally found in the environment and when these concentrations exceed a stipulated limit, they may become toxic to the surrounding environment [2]. From an environmental point of view, coastal zones can be considered as the geographic space of interaction between terrestrial and marine species. The coastal zones are received a large amount of metal pollution from agricultural runoff, aquaculture chemicals and other industrial activities. Adverse anthropogenic effects on the coastal environment include eutrophrication, heavy metals, organic and microbial pollution, and port activities. The discharge of these wastes without adequate treatment often contaminate the estuarine and coastal waters with conservative pollutants (like heavy metals), may of which accumulate in the tissues of the resident organisms like fishes and other aquatic organisms

Fish, as human food, are considered source of protein, polyunsaturated fatty acids particularly omega-3 fatty acids, Calcium, Zinc and Iron [3]. And it is considered one of the high nutrient sources for humans that contribute the lower the blood cholesterol and reduce the risk of stroke and heart diseases [4,5]. Among the aquatic fauna, fish is most susceptible to heavy metal contamination than any other aquatic fauna. It is well known that fish are good indicator of chemical pollution and as a result they long been used to monitor metal pollution in coastal and marine environment. So, fishes were considered as better specimens for use in the investigation of pollution load than the water sample because of the significant levels of metals they bioaccumulate. Hence, harmful substances like heavy metals, released by anthropogenic activities will be accumulated in marine organisms through the food chain; as result, human health can be at risk because of consumption of fish contaminated by toxic chemicals.

Keeping inview of the potential toxicity, persistent nature, as well as the environmental pollution, it is deemed necessary to have the base line environmental data on potential metal contamination so that pollutants can be judged in the environment. This paper presents the data on heavy metal (Zn, Pb, Mn, Cu, Cr and Hg) concentration in fish, *Mugil cephalus* from Machilipatnam coast.

2. MATERIALS AND METHODS

Water and fish samples collected from fish landing centre, Machilipatnam (Lat. 16° 11′ 01 N and Long. 81° 10′ 42.3 E). The fish samples transported to the laboratory in ice boxes and stored at -10°C until subjected for future analysis. The fishes were dissected and care was taken to avoid external contaminated to the samples. Rust free stainless steel kit was sterilized to dissect the fishes. Double distilled water was used for making up the sample and for analysis in the Atomic Absorption Spectrophotometer (ASS). The gut content, gill and muscles were separated and dried to constant weight and both wet and dry weight recorded. 25% was used as blank samples accompanied every run of the analysis. Each sample was analyzed in triple to ensure accuracy and precession for the analytical procedure.

2.1 Health Risk Assessment

Estimated daily intake (EDI):

$$EDI = \frac{E_F \times E_D \times F_{IR} \times C_f \times C_m}{W_{AB} \times T_A} \times 10^{-3}$$

E_F = The exposure frequency 365 days/year

E_D = The exposure duration, equalent to average life time (65 years)

F_{IR} = The fresh food ingestion rate (g/person/day) which is considered to be India 55/g/person/day [6].

C_f = The conversion factor = 0.208

C_m = The heavy metal concentration in food stuffs mg/kg d-w)

W_{AB} = average body weight (bw) (average body weight to be 60kg)

TA = Is the average exposure of time for non carcinogens (It is equal to ($E_F \times E_D$) as used by in many previous studies [7].

Target hazard quotient:

$$THQ = \frac{EDI}{RfD}$$

Rfd: Oral reference dose (mg/kg bw/day)

A THQ below 1 means the exposed population is unlikely to experience obviously adverse effects, whereas a THQ above means that there is a chance of non-carcinogenic effects, with an increasing probability as the value increases.

3. RESULTS AND DISCUSSION

The purpose of this work to determined the presence of a particular group of metals in the water ecosystem of the Nizampatnam harbor area. Heaving record to the possibility of bioaccumulation of these metals in tissues of living organisms, including fish it was necessary to find out whether the metals determined in the water samples were to be accumulated in the fish fillet (Edible parts), the risk imposed on a local population was evaluated. The research presented herein had been conducted in the determination heavy metals concentration in fish fillet (Muscle) sample.

3.1 Heavy Metals in Fishes

The mean concentrations of heavy metal in fish muscle are presented Table 1, Fig. 1. The order of heavy metal concentration was Zn>Pb>Mn>Cu>Cr>Hg. This data indicated Zinc accumulated.

3.1.1 Zinc (Zn)

Zinc is an essential element in animal's diet but it is regarded as potential hazard for both animals and human health [8]. Insignificant seasonal variation is observed with slight higher concentration during monsoon season.

Table 1. Average (Mn, Pb, Cu, Zn, Hg and Cr) concentrations in Liver and Muscle of fish collected from Machilipatnam coast (mg/kg dry weight)

Fish	No.	Mn			Pb			Cu			Zn			Hg			Cr		
		M	L	A	M	L	A	M	L	A	M	L	A	M	L	A	M	L	A
M. cephalus	60	6.3	11.5	8.9	8.4	13.2	10.8	5.5	7.3	6.4	25.2	39.6	32.4	1.5	2.9	2.2	1.6	3.0	2.3

No.: Number; M: Muscle; L: Liver; A: Average

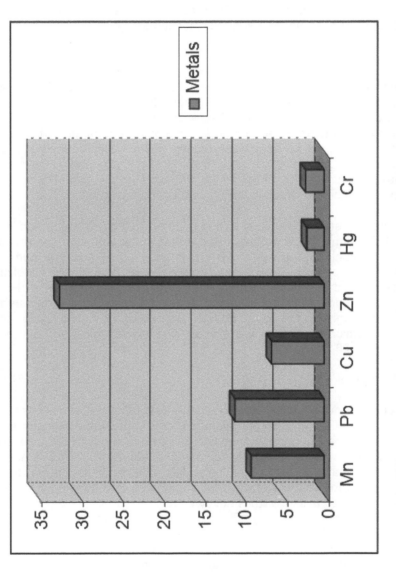

Fig. 1. Different concentrations of heavy metals in fish (mg/kg, dry weight)

Zinc is present in natural water only as a miner consultant because lack of solubility of free metal and its oxides [9]. It is a very high concentration only it may causes some toxic effects. A normal human body contains 1.4 to 2.3 g of Zinc. Recommend daily dietary intake of Zinc is about 15mg for adults and 10mg for children over a year old. The average diary intake of zinc in India is about 16.1mg [10]. It is relatively non toxic and concentrations of Zinc up to 25mg/l have shown few adverse effects [11]. Zinc may be toxic aquatic organisms but the degree of toxicity varies greatly, depending on water quality characteristics as well as species being considered [12]. The present study shows that the average concentration 32mg/kg of Zn much higher than WHO standards [13].

3.1.2 Lead (Pb)

Pb is considered as a toxic but non-essential metal implying that it has no known function in the biochemical processes [14]. Lead enters the aquatic environment through soil erosion and leaching gasoline combustion, municipal and industrial wastes and runoff [15]. Pregnant women exposed to lead were found to have high rates of still births and miscarriages [16]. Lead has caused mental retardation among children. Hyper tension caused by Pb exposure has also been reported [17]. Lead poising is a accompanied by symptoms of intestinal cramps, peripheral nerve paralysis anemia, and fatiage [18]. The concentration of lead in natural water increases mainly through anthropogenic activities [19].

In the present study Pb concentration goes to 10.8mg/kg in the fish muscle. According to WHO [13], the maximum accepted limit is 2mg/kg for food fish. The present results indicated that the concentration levels of Pb was mostly higher than the permissible limits set for human consumption by various regulatory agencies and therefore indicated possible health risks associated with consumption of these fish. At high levels of Pb exposure these is damaged to almost all organ systems. Most importantly the central nervous system kidneys, and blood, culminating in death, if levels are excessive. At low levels, haeme synthesis and other biochemical processes are affected and psychological and neurobehavioral functions are impaired [20,21].

3.1.3 Manganese (Mn)

Manganese is an essential micro nutrient, as it functions as a co factor for many enzyme activities [22]. High Mn concentration interferes with central nervous system of vertebrates by inhibiting dopamine formation as well as interfere ring with other metabolic pathways such as Na regulation which ultimately can cause death. High Mn levels are a matter of concern as the consumption of Mn contaminated fish could result in the Mn related disorders in the consumers. In the present study manganese goes to 8.9 mg/kg in the fish muscle which is higher than the permissible limits set by WHO [13]

3.1.4 Copper (Cu)

Copper in aqueous systems received attention mostly because of its toxic effects on biota. Excess of Cu in human body is toxic and hypertension and causes some disorders. Cu also produces pathological changes in brain tissues [23]. The average concentration of Cu in the present study goes to 6.4mg/kg in fish muscle who is above permissible limits.

3.1.5 Chromium (Cr)

Chromium concentration in natural waters is usually low. Elevated concentration can result from industrial and mining processes (12). Fish are usually more resistant to Cr than other aquatic organisms, but they can be affected sub-lethally where exposed to concentration increases. In the present study Cr also above permissible levels set by WHO [13]

3.1.6 Mercury (Hg)

Mercury is known to be latent neurotoxin compared to other metal like lead, cadmium, copper. A high dietary intake of mercury from consumption of fish and fishery has been hypothesized increase the risk of coronary heart disease [24]. When deposited in biota, mercury undergoes biotransformation, in which organic mercury (methyl/ mercury). Microbes subsequently concentrate mercury through the food chain in the tissues of fish and marine animals [25]. According to results obtained, the mercury levels of muscle of M. cephalus were found to be 2.2 mg/kg which was higher than permissible levels of WHO [13]. Data we found well the expected levels of concentration in the sample area in relation to the national and international contributions and to those of anthropogenic origin. The last few decades were witness to several reports on the toxicity of heavy metals in human beings, due to the contamination in aquatic organisms. Predominantly, fish toxicological and environmental studies have prompted interest in the development of toxic elements in sea food [26].

The increasing demand of food safety has accelerated researching regarding the risk associated with food consumption contaminated by heavy metal [27]. Lon term intake of contaminated sea food could lead to toxicity of heavy metals in human beings. There are reports of high levels of heavy metals are natural components of food stuffs but also because of environmental contamination and contamination during processing [28]. Industrial effluents agriculture runoff, aquaculture chemicals and drugs, animal and human excretion, and geological weathering and domestic waste contribute to the heavy metal in the water bodies [29]. With the exception of occupational exposure, fish are acknowledge to be single largest source of mercury and other heavy metals (lead and chromium) affecting human beings. Lead poisoning in children causes neurological damage leading to reduced intelligence, loss of short-term memory, learning disabilities and coordination problems. The threat of heavy metal to human and animal health is aggravated by their long-term persistence in the environment [30].

Further, the heavy metals causing concern is that they may be transferred and accumulated in the bodies of animals or human beings through food chain, which will probably cause DNA damage and carcinogenic effects due to their mutagenic ability [31]. Heavy metal exposure of the population may cause neurobehavioral disorders. Such as fatigue insomnia decreased concentration, depression, irritability, sensory and motor symptoms [32]. Exposure to heavy metals has been linked to developmental retardation, various types of cancer, kidney damage, autoimmunity and even death in some instances of exposure to very high concentrations [33]. In some cases fish catches were banned for human consumption because their heavy metal concentrations exceeded the maximum limits recommended by the Food and Agriculture organization (FAO) and world health organization (WHO). Among sea foods, fish are commonly consumed and hence, are a connecting link for the transfer of toxic heavy metals in human beings. Bhuvaneshwari et al. [34] concluded that the metals are an inherent component of the environment that pose a potential hazard to human beings and animals. The consumption of fish from the polluted site may result in accumulation of persistent pollutants in ultimate recent of food web. The effluents from the textile factory, the tannery and the floriculture farm probably contain harmful contaminants such as dye stuffs, bonzothiozole, sulphonated polyphenols and pesticides. These compounds could bioaccumulate and affect the health of aquatic organisms and subsequently, the health of humans, as consumers of these fish [35]. Türkmen et al. [36] reported that metals in tissues of fish species from Akyatan Lagoon. Türkmen et al. [37] observed that the metals in tissues of fish from Paradeniz Lagoon in the Coastal Area of Northern East Mediterranean. Türkmen et al. [38] worked on heavy metal levels in Blue Crab (*Callinectes sapidus)* and Mullet (*Mugil cephalus*) in İskenderun Bay (North Eastern Mediterranean, Turkey). In the present study Machilipatnam coast also effected pollutants particularly dyes factory, agriculture and aquaculture chemicals.

The fish, we analysed revel some metals concentrations potentially toxic if they enter the food chain. However, since their toxicity for human is given by the ingestion rate, data obtained on THQs values (Mn-4.6; Pb-5.6; Cu-3.3; Zn-16.8; Hg-1.15; Cr-1.2) indicated that the contractions we found in the sample of fish represent a risk for human health because all metals THQ is higher than one. Of course, it is just a Primary step; fish contamination levels should be carefully monitored on a regular basis, to detect any change in their patterns that could become a hazard on human safety. Similar results observed by Ambedkar and Maniyan [39]. They concluded that the heavy metal concentrations were above the maximum levels recommended by regulatory agencies and, depending on daily intake by consumers, might represent a risk for human health.

4. CONCLUSION

The international official regulatory agencies like WHO have set limits for heavy metal contaminations above which the fish and fishery products are unsuitable for human consumption. However, in the Indian subcontinent there is no safety levels of heavy metal in fish tissues although the Indian population is the major fish consumers in the tropics with a weekly annual rate of 55kg/person.

Finally, we recommended that a long-term continuous monitoring to check metals pollution, in order to control of metal in water and fish, control and assessment of the metal content in water of Machilipatnam area which are supplied by water used agriculture, aquaculture, industries (particularly dies factories), quality of water farmlands. And also quality control of input and output water into coastal zones in Machilipatnam area has widely importance. In addition, guidance of people and farmers of both agriculture and aquaculture, about the instruction for use of pesticides, chemicals, drugs and control of house wastewater spreading in rivers and crops are necessary.

COMPETING INTERESTS

Authors have declared that no competing interests exist.

REFERENCES

1.   Jordao CP, Pereira MG, Bellato CR, Pereira JL, Matos AT. Assessment of water system for contaminants from domestic and industrial sewages, Environment Monitoring and Assessment. 2002;79:75-100.

2.   WHO. Inorganic Mercury Environmental Health Criteria. Vol. 118. Geneva- World Health Organization; 1591-1991.

Heavy Metals Concentration in Fish Mugil cephalus from Machilipatnam Coast and Possible...

31

3.   Chan HM, Tritonopoules M, Img A, Receeveur O, Johnson E. Consumption of fresh water fish in Kahn awake: risks and benefits. Environ. Res. Sec. 1999;80(2):213-22.

4.   Storelli MM, Potential human health risk from metal (Hg, Cd and Pb) and polychlorinated biphenyl's (PCBC) via sea food consumption: Estimation of target hazard quotients (THQs) and toxic equivalent (TEOs). Food and Chemical Toxicology. 2008;46:2782-8.

5.   Al-Busaidi M, Yesudhason P, Al-Mughairi S, Al-Rahbi WAK, Al-Harthy KS, Al-Mazrooei NA et al. Toxic metals in commercial marine fish in Oman with reference of national and international standards. Chemosphere. 2011;85:67-73.

6.   Mitra A. Chowdhury R, Benerjee K. Concentration of some heavy metal in commercially important fin fish and shell fish of the River Ganga. Environ. Monit. Assess. 2012;184:2219-30.

7.   Wang X, Santo T, Xing B, Tao S. Health risk of heavy metals to the general public in Tianjin, China via consumption of vegetable and fish. Science of the Total Environment. 2005;350:28-37.

8.   Amundsen PA, Staldvilk FJ, Lukin A, Kashulin N., Popova O, Restetnikov Y. Heavy metals contaminations in fresh water fish form the border region between Norway and Russia. Science of the Total Environment. 1997;201:211-4.

9.   David E, Irvin G, Knights B. Pollution and the use of chemicals in agriculture. Butter worth, London.

10.   Krishnan K. Fundamentals of environmental pollution. Chand. Ltd. 1995;366.

11.   McNeely RN, Neimanis VP, Dwyer L. Water quality source book. A guide to water quality parameter, Inland water directorate (pp.1-65). Water quality Branch, Ohawa. Canada.

12.   Datar MD, Vashishtha RP. Investigation of heavy metals in water and silt sediments of Betwa River. Indian Journal of Environmental Protection. 1990;10(9):666-72.

13.   WHO (World health organization). Guidelines for drinking water quality. Recommendation WHO. Geneva. 1985;1:130.

14.   Adeyeye EI, Akinyugha NJ, Fesobi ME, Tenabe VO. Determination of some metals in freshwater pond and their environment. Aquaculture. 1996;47:205-14.

15.   DWAF (Department of water Affairs and Forestry). South African water quality guidelines, aquatic ecosystems. 2nd Ed. 1996;7:159.

16.   WHO. Technical report series: Trace elements in human nutrition. Report of a WHO expert committee. No.532. Geneva; 1973.

17.   Beevens DG, Erskine E, Robertson M, Beattle AD, Campbell BC, Goldberg A. Blood lead to hypertension. Lancet. 1976;2:1-3.

18.   Umar A, Umar R, Ahmad MS. Hydrogeological and hydrochemical frame works of regional aquter system in Kali-Ganga sub-basin. India. Environmental Geology. 2001;40(4-5):602-11.

19.   Goel PK. Water pollution causes effects and control. New Delhi, New age; 1997.

20.   Goldstein GW. Neurological concept of lead poisoning in children. Pediatr Ann. 1992;21:384-8.

21.   World Health Organization. Inorganic lead, Environmental Health Criteria. No.165. World Health Organization, Geneva, Switzerland; 1995.

22.   Suresh B, Steiner W, Rydlo M, Taraschewski H. Concentration of 17 elements in zebra mussel (Dreissena polymorpha). Environmental Toxicology and Chemist. 1999;18:2574-9.

23. Kudesia VP. Water pollution. Merrut. Pragati Prakasam; 1990.

24. Salonen JT, Seppanen K, Nyyssonen K, Korpela H, Kauhanen J, Kantola M, et al. Intake of mercury from fish, lipid peroxidation, and the risk of myocardial infarction and coronary, cardiovascular, and any death in eastern Finnish men. Circulation. 1995;91:645-55.

25. Altindag A, Yigit, S. Assessment of heavy metal concentrations in the food web of lake Beysehir, Turkey. Chemosphere. 2005;60(4):552-6.

26. Waqar A. Levels of selected heavy metals in tuna fish. The Arabian Journal for Science and Engineering. 2006;31:89-92.

27. Mansour SA, Belal MH, Abou-Arab AAK, Gad MF. Monitoring of pesticides and heavy metals in cucumber fruits produced from different farming systems. Chemosphere. 2009;75(5):601–9.

28. Kalay M, Aly O, Canil M. Heavy metal concentration in fish tissues from the Northeast Mediterranean Sea. Bulletin of Environmental contamination and Toxicology. 1999;63:673-81.

29. Erdogrul O, Erbilir F. Heavy metal and trace elements in various fish samples from Sir Dam Lake, Kahramanamras, Turkey. Environmental Monitoring and Assessment. 2007;130:373-9.

30. Gisbert C, Ros R, de Haro A., Walker DJ, Bernal MP, Serrano R, et al. A plant genetically modified that accumulates Pb is especially promising for phytoremediation. Biochemical and Biophysical Research Communications. 2000;303:440-5.

31. Knasmuller S, Gottmann E, Steinkellner H, Fomin A, Pickl C, Paschke A, et al. Detection of genotoxic effects of heavy metal contaminated soils with plant bioassay. Mutation Research. 1998;420:37-48.

32. Hanninen H, Lindstrom H. Behaviour test battery for toxic psychological studies used at the institute of occupational health in Helsinki. Helsinki: Institute of Occupational Health; 1979.

33. Rai PK. Heavy metal phytoremediation from aquatic ecosystems with special reference to macrophytes. Critical Reviews in Environmental Science and Technology. 2009;39(9):697-753.

34. Bhuvaneshwari R, Mamtha N, Selvam P, Rajendran RB. Bioaccumulation of metals in muscle, liver and gills of six commercial fish species at Anaikarai Dam of River Kaveri, south India. International Journal of Applied Biology and Pharmaceutical Technology. 2012;3:8-12.

35. Dsikowitzky L, Mengesha M, Dadebo E, de Carvalho CEV, Sindern S. Assessment of heavy metals in water samples and tissues of edible fish species from Awassa and Koka Rift valley lakes, Ethiopia. Environ. Monit. Assess. 2013;185:3117-31.

36. Türkmen A, Tepe Y, Türkmen M, Ateş A. Investigation of metals in tissues of fish species from Akyatan Lagoon. Fresenius Environmental Bulletin. 2012;21(11c):3562-7.

37. Türkmen M, Türkmen A, Tepe Y. Comparison of metals in tissues of fish from Paradeniz Lagoon in the coastal area of Northern East Mediterranean. Bulletin of Environmental Contamination and Toxicology. 2011;87(4):381-5.

38. Türkmen A, Türkmen M, Tepe Y, Mazlum Y, Oymael S. Heavy metal levels in blue crab (*Callinectes sapidus*) and mullet (*Mugil cephalus*) in İskenderun Bay (North Eastern Mediterranean, Turkey). Bulletin of Environmental Contamination and Toxicology. 2006;77(2):186-93.

39. Ambedkar G, Muniyan M. Bioaccumulation of some heavy metals in the selected five freshwater fish from Kollidam River, Tamil Nadu, India. Advances in Applied Science Research. 2011;2(5):221-5.

Listeria spp. in Raw Cow and Goat Meat in Port Harcourt, Nigeria

O. C Eruteya[1*], S. A Odunfa[2] and J. Lahor[1]

[1]*Microbiology Department, University of Port Harcourt, Port Harcourt, Nigeria.*
[2]*Department of Microbiology, University of Ibadan, Ibadan, Nigeria.*

Authors' contributions

Author SAO supervised the research; read and approved the manuscript. Author OCE designed the study, wrote the draft, participated in the isolation, confirmation and handled the molecular aspect. Author JL handled the purchasing of samples, isolation and confirmation of isolates.

ABSTRACT

Aims: To determine the occurrence of *Listeria monocytogenes* and other *Listeria* spp. in raw cow and goat meat.
Study Design: This work was based on a completely randomized design with two replication and the average values calculated for mean comparison.
Place and Duration of Study: Department of Microbiology, University of Port Harcourt, Nigeria and Nigeria Institute of Medical Research, Yaba, Lagos, Nigeria. Isolation of *Listeria* spp. was between March 2011 and February 2012.
Methodology: In this study, a total of 240 raw cow and goat meat samples were analyzed for the presence of *Listeria monocytogenes* and other *Listeria* species. The techniques recommended by the United State Department of Agriculture (USDA) revised and the Health Products and Food Methods of the Government of Canada were employed using Fraser broth and polymixin acriflavin lithium chloride ceftazidime aesculin mannitol (PALCAM) agar.
Results: The results of conventional and polymerase chain reaction (PCR) characterization of the isolates revealed that 81 samples (33.75%) were positive for *Listeria* spp. The highest prevalence of *Listeria* was found in cow flesh samples (19 of 36 samples; 52.78%) followed by cow intestine samples (8 of 20; 40%) and least was goat kidney samples (5 of 28; 17.86%). Out of the 310 characterized *Listeria* spp., *L.*

*Corresponding author: Email: onoriode.eruteya@uniport.edu.ng, chrisruteya@yahoo.com *monocytogenes* were 4(1.29%). Other species isolated were *L.* innocua 20(6.45%), *L. ivanovii* 4(1.29%), *L. seeligeri* 72 (23.23%), *L. welshimeri* 139 (44.84%) and *L. grayi* 71(22.90%). No *L. monocyogenes* was isolated from cow samples. A higher incidence was noted during the raining season 216 (69.68%) than the dry season 94(30.32%). **Conclusion:** This study demonstrated the occurrence and distribution of *Listeria* species in retail raw cow and goat meat in Port Harcourt, Nigeria.

Keywords: Cow meat; goat meat; listeria; listeriosis; PCR.

1. INTRODUCTION

Meat is the major source of protein and valuable qualities of vitamins for most people in many parts of the world, thus they are essential for the growth, repair and maintenance of body cells and necessary for our everyday activities [1]. Due to the chemical composition and biological characteristics, meats are highly perishable foods which provide excellent source for growth of many hazardous microorganisms that can cause infection in humans and spoilage of meat and economic loss [2]. Listeric infections, caused by microorganisms of the genius *Listeria*, occur worldwide and in a variety of animals including man [3]. Cases of listeriosis arise mainly from the ingestion of contaminated food and the disease is particularly common in ruminants fed on silage [3]. The genus *Listeria* comprises six species: *L. monocytogenes, L. innocua, L. welshimeri, L. grayi, L. ivanovii* and *L. seeligeri* [4]. Two of these species, *L. monocytogenes* and *L. ivanovii*, are potentially pathogenic [5]. The other four *Listeria* species are essentially saprophytes that have adapted for survival in soil and decaying vegetation [4]. *Listeria monocytogenes*, a facultative intracellular pathogen, repeatedly found in meat and meat products, raw milk, soft cheese and pasteurized dairy products, vegetables, and fish and fish products [6] is responsible for severe foodborne infections in humans of all ages but especially pregnant women, infants less than four weeks old, the elderly and immunocompromised individuals and also cause invasive disease in many ruminants including, cattle, sheep, and goats species [7-9]. The bacterium possesses properties that favor it as a foodborne pathogen: at variance with most other pathogens it is relatively resistant to acid and high salt concentrations; it grows at low temperature, down to freezing point, which mean it may grow in refrigerated foods, measures commonly used to control the growth of pathogens in foods [10,11]. Before the development of *Listeria* selective media, this property was used for selective enrichment of the bacterium from complex matrices, with *Listeria* outnumbering the competing flora after incubation of the enrichment culture at refrigerator temperature for weeks or months [11].

Although the occurrence of *Listeria* in various foods has been investigated in several countries, there is limited information regarding its prevalence in raw meat and foods in Nigeria. The present study was conducted to determine the occurrence of *Listeria* species in raw cow and goat meat in Port Harcourt, Nigeria.

2. MATERIALS AND METHODS

2.1 Sample Collection

A total of 240 raw cow and goat meat samples comprising 122 cow and 118 goat meat were randomly purchased from meat vendors in three strategic markets in Port Harcourt metropolis from March, 2011 to February, 2012. The samples were kept in ice box containing ice packs and immediately transported to the Microbiology laboratory, University of Port Harcourt where they were kept in refrigerator until analyzed.

2.2 Isolation of Listeria sp

The techniques recommended by the United State Department of Agriculture (USDA) revised [12] and the Health Products and Food Methods of the Government of Canada [13] were employed using Fraser broth (Oxoid, England) and polymixin acriflavin lithium chloride ceftazidime aesculin mannitol (PALCAM) agar (Oxoid, England). Twenty-five grams of each meat samples was added to a stomacher bag containing 225ml of sterile half-Fraser broth and supplements. The mixture was homogenized using a stomacher (lab-blender, Seward medical, London) at high speed for 1-2min. The test portion was incubated at $30^{\circ}C$ for 24h. From the pre-enrichment culture (half Fraser broth), 0.1ml was transferred into 10ml of full-strength Fraser broth with supplements added and was incubated at $35^{\circ}C$ for 24-48h. From the culture obtained in Fraser broth showing evidence of darkening due to aesculin hydrolysis by *Listeria* sp, 0.1ml was transferred onto duplicate PALCAM plates. After spreading, plates were incubated at $37^{\circ}C$ for 24- 48h. The plates were examined for the presence of characteristic colonies presumed to be *Listeria* sp- 2mm grey-green colonies with a black sunken centre and a black halo on a cherry-red background, following aesculin hydrolysis and mannitol fermentation. Five typical colonies were selected randomly from a pair of PALCAM plates for confirmation and subsequent identification.

2.3 Confirmation and Identification

Colonies suspected to be *Listeria* were transferred onto trypticase soy agar (Becton, Dickinson and company, France) with 0.6% yeast extract (Lab M, UK) and incubated at $37^{\circ}C$ for 18 to 24h, before being subjected to the following standard biochemical tests: gram staining, catalase reaction, oxidase reaction, beta haemolysis on sheep blood agar and acid production from mannitol, rhamnose and xylose. Confirmed isolates on the basis of criteria suggested by Seeliger and Jones [14] were further identified using the polymerase chain reaction (PCR).

2.4 Extraction of *Listeria* DNA (Deoxyribonucleic acid)

DNA was extracted by the boiling method without triton x-100 [15]. Cells were harvested by centrifugation (Eppendorf centrifuge 5418, Germany) of overnight brain heart infusion broth culture of *Listeria* in 2ml eppendorf tube at 10,000rpm for 2min and the supernatants discarded. The pellets were re-suspended in 1ml sterile distilled water and centrifuged after vortexing (Vortexer 59a, Denville Scientific Inc, Taiwan) at 10,000rpm for 5min. The supernatants were again discarded and the pellets resuspended in 200μl sterile water and vortexed. The suspensions were heated for 10min in a boiling bath ($100^{\circ}C$) (Grant GLS400, Grant Instrument, England). After cooling and vortexing, the mixtures were centrifuged at 10,000rpm for 5min. The supernatants were then transferred to a prelabelled 1.5ml eppendorf tube while the sediments were discarded. The DNA extracted was stored in deep freezer ($-20^{\circ}C$) until further analysis.

2.5 Identification by PCR

Oligonucleotide primers described by Border et al. [16] for all *Listeria* species (U1[5'-CAGCMGCCGCGGTAATWC-3'] and LI1[5'-CTCCATAAAGGTGACCCT-3']) and for all serotypes of *L. monocytogenes* (LM1[5'-CTAAGACGCCAATCGAA-3'] and LM2 [5'-

AAGCGCTTGCAACTGCTC-3']) and Bubert et al. [17] for all *Listeria* species (primers MonoA [5'- TTATACGCGACCGAAGCCAAC-3'] and Lis1B [5'-TTATACGCGACCGAAGCCAAC-3']), all serotypes of *L. innocua* (Ino2 [5'-ACTAGCACTCCAGTTGTTAAAC-3'] and Lis1B), all serotypes of *L. grayi* (Mugra1[5'-CCAGCAGTTTCTAAACCTGCT-3'] and Lis1B),all serotypes of *L. welshimeri* (Wel1 [5'-CCCTACTGCTCCAAAAGCAGCG -3'] and Lis1B), all serotypes of *L. ivanovii* (Iva1 [5'-CTACTCAAGCGCAAGCGGCAC -3'] and Lis 1B), all serotypes of *L. seeligeri* (Sel1 [5'-TACACAAGCGGCTCCTGCTCAAC -3'] and Lis1B) while the grouped species, *L. seeligeri, L. ivanovii* and *L. welshimeri* were identified using Siwi2 (5'- TAACTGAGGTAGCGAGCGAA -3') and Lis1B, synthesized by Biomers.Net GMBH, Germany were employed.

The reactions involving U1, LI1, LM1 and LM2 were carried out in a final volume of 25µl, containing 2.5 µl 10×PCR buffer, 1.5 µl MgCl$_2$, 0.5µl dNTP (deoxynucleoside triphosphate), 0.25 µl each of appropriate primer, 0.1 µl AmphiTaq DNA polymerase(all products of Solis BioDyne, Estonia),1.5µl of appropriate DNA preparation and 18.4µl sterile distilled water. amplification following an initial denaturation at 95°C for 3min was performed in 30 cycles, at 95°C for 30s, 50°C for 60s and 72°C for 60s in a thermo cycler (Mastercycler-Eppendorf, Vapo-product,Germany). A final extension was performed for 10min at 72°C. A 8µl aliquot of PCR product mixed with a loading dye (10mm, EDTA, 10% glycerol, 0.015% bromo phenol blue and 0.017% sodium dodecyl sulphate(SDS), made up to 100ml) were checked in an ethidium bromide stained 1.5% agarose(Fermentas, Life Science, Germany) and the gel read in a UV transilluminator (Genosens 1500, Clinx Science Instruments Co. Ltd, China). Reaction mixture with the DNA of *L. monocytogenes* PCM 2191 serovar 01/2 (Polish Collection of Microorganisms, Poland) template serve as positive control while a reaction mixture with no DNA template was incorporated as a negative control in each reaction.

The reactions involving MonoA, Iva1, Sel1, Wel1, Ino2, Mugra1, Siwi2, Lis1A and Lis1B were also carried out in a final volume of 25µl, containing 2.5 µl 10×pcr buffer, 1.5 µl MgCl$_2$, 0.5µl DNTP (deoxynucleoside triphosphate), 0.2 µl each of appropriate primer, 0.15 µl AmphiTaq dna polymerase (all products of Solis BioDyne, Estonia), 1.5µl of appropriate DNA preparation and 18.45µl sterile distilled water. amplification following an initial denaturation at 95°C for 3min was performed in 30 cycles, at 95°C for 30s, 58°C for 60s and 72°C for 60s in a thermo cycler (Mastercycler-Eppendorf, Vapo-product,Germany). A final extension was performed for 10min at 72°C. a 8µl aliquot of pcr product mixed with a loading dye (10mm, EDTA, 10% glycerol, 0.015% bromo phenol blue and 0.017% sodium dodecyl sulphate (SDS), made up to 100ml) were checked in an ethidium bromide stained 1.5% agarose (Fermentas, Life Science, Germany) and the gel read in a UV transilluminator (Genosens 1500, Clinx Science Instruments Co. Ltd, China). Reaction mixture with the DNA of *L. monocytogenes* PCM 2191 serovar 01/2 (Polish Collection of Microorganisms, Poland) template serve as positive control while a reaction mixture with no DNA template was incorporated as a negative control in each reaction.

2.6 Statistical Analysis

The distribution of the *Listeria* species in the various meat types and parts were subjected to analysis of variance (ANOVA) and Duncan [18] to determine means that differed.

3. RESULTS AND DISCUSSION

Out of a total of 240 meat samples analyzed, 81(33.75%) were positive for *Listeria*. *Listeria* species were isolated from all the beef and goat meat parts examined (Table 1). The level of contamination of meat samples by *Listeria monocytogenes* and other *Listeria* species varied and was highest in cow flesh (19 of 36 samples, 52.78%), followed by cow intestine (8 of 20 samples, 40%) and least was goat kidney (5 of 28 samples, 17.86%).

Table 1. Detection of *Listeria* species in raw cow and goat meat

Meat type	Number of samples		%
	Examined	Positive	
Cow flesh	36	19	52.78
Cow intestine	20	8	40.00
Cow kidney	32	9	28.13
Cow liver	34	12	35.29
Goat flesh	36	11	30.56
Goat intestine	18	5	27.78
Goat kidney	28	5	17.86
Goat liver	36	12	33.33
Total	240	81	33.75

out of the 310 characterized *Listeria,* 74(23.87%) were from cow flesh, 29(9.35%) from cow intestine, 41(13.23%) from cow kidney, 46(14.84%) from cow liver, 33(10.65%) from goat flesh, 19(6.13%) from goat intestine, 17(5.48%) from goat kidney and 51(16.45%) from goat liver (Table 2).

Table 2. Distribution of *Listeria* species isolated from various meat types/parts.

Listeria sp.	Number of *Listeria* species isolated from each meat type/parts								Total
	CF	CL	CK	CL	GF	GI	GK	GL	
L. monocytogenes	0	0	0	0	3	1	0	0	4
L. innocua	5	3	1	3	1	0	1	6	20
L. seeligeri	18	1	13	8	9	4	7	12	72
L. ivanovii	2	0	1	1	0	0	0	0	4
L. grayi	14	8	10	11	9	6	1	12	71
L. welshimeri	35	17	16	23	11	8	8	21	139
Total	74	29	41	46	33	19	17	51	310

CF= cow flesh; CI= cow intestine; CK= cow kidney; CL= cow liver; GF= goat flesh; GI= goat intestine; GK= goat kidney; GL= goat liver.

All 310 *Listeria* isolates identified as *L. monocytogenes* 4(1.29%), *L. innocua* 20(6.45%), *L. seeligeri* 72(23.23%), *L. ivanovii* 4 (1.29%), *L. welshimeri* (139%) and *L. grayi* 71(22.90%) by conventional method were also positive using the PCR assay (Fig. 1). There were no significant difference (p=.05) in the occurrence of *Listeria* among the different meat parts examined.

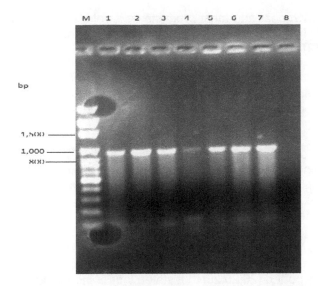

Fig. 1. Products obtained when total genomic DNA from the reference
***L. monocytogenes* PCM 2191serovar 01/2 strain and representative of each isolated**
***Listeria* species were subjected to PCR using LI1 and U1 primer combination for all**
***Listeria* species. A PCR product of 938bp was observed. Lanes: m, molecular weight**
standard; 1, *L. innocua*; 2, *L. welshimeri*; 3 *L. seeligeri*; 4, *L. grayi*; 5, *L. ivanovii*; 6, *L.*
***monocytogenes*; 7, *L. monocytogenes* PCM 2191serovar 01/2 (positive control); 8,**
control reaction (all reagent ingredients except chromosomal DNA). PCR products
were separated in a 1.5% agarose gel and stained with ethidium bromide.

The highest occurrence of *Listeria* was in July (15.81%); it was also high in August (12.58%), both being the peak of the raining season in Port Harcourt (Fig. 2). Overall, a higher incidence was noted during the raining season 216 (69.68%) than the dry season 94(30.32%).

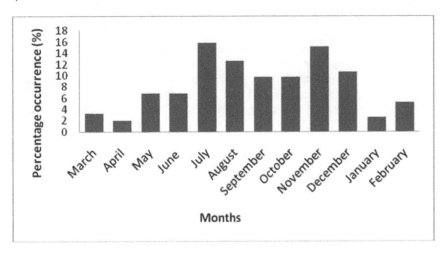

Fig. 2. Monthly Distribution of Isolated *Listeria* Species

The importance of raw meat and meat products as a vehicle for the transmission of various diseases, especially in countries where hygienic standards are not strictly enforced has been well documented [19]

It is not surprising that *L. monocytogenes* and other *Listeria* spp. were found in meat considering its widespread occurrence in nature, especially in plants, and their associations in nature with other bacteria that are well established in meats make it reasonable to find it in meats [20].

Overall, 33.75% (81 of 240) of all meat samples in this study were contaminated with *Listeria* species. This incidence of *Listeria* in the meat examined coincides with the report of a 0 to 68% prevalence of *Listeria* in fresh meat [21]. This can be attributed to the consumption of contaminated silage or other feeds, fecal contamination during evisceration and from handlers and slaughterhouses [3,22-24].

In this study, *L. welshimeri* was the predominant *Listeria* species isolated from the beef and goat meat samples; other *Listeria* species were less common (Table 2). Our result differed from those already reported [20,25,26] that stated that *L. monocytogenes* and *L. innocua* are most often reported with most investigators reporting one or the other to be most predominant and *L. welshimeri* coming third or next after *L. innocua*, followed by *L. seeligeri* and lastly *L. ivanovii* in meat among this five related species. *L. monocytogenes* accounted for 4(1.29%) of the 310 *Listeria* isolated. None was from the beef samples. Previous studies in Nigeria, India, Serbia and Bangkok, reported the inability to isolate *L. monocytogenes* in raw-goat, -beef and – meat [22,27-29]. However, a number of authors reported a 4.65% and 6.4% (Bulgaria), 5.1% (Ethiopia), 6.66% (India), 17.7% (Portugal), 20% (Greek), 31% (Denmark) and 35% (Spain) prevalence of *L. monocytogenes* from raw- beef, -goat and– meat [10,30-35], making obvious that the occurrence of *L. monocytogenes* varies from one place and one author to the other.

The distribution of the *Listeria* species (Fig. 2) shows that the highest occurrence was in July (15.81%). The occurrence was also high for August (12.58%), both being the peak of the raining season in Port Harcourt, Nigeria. This is comparable with a high incidence of *Listeria* in beef and goat meat during fall in Iran [36].

The inability of the Lis1A and Lis1B and MonoA and Lis1B primer combinations to produce the expected PCR product for all *Listeria* species and *L. monocytogenes* respectively, occasioned the use of primer combination; Ll1 and U1 (Fig. 1) and LM1 and LM2 targeting the more conserve genomes of all *Listeria* and *L. monocytogenes* respectively. However, the others produced the expected PCR product, amongst them Siwi2 and Lis1B (Fig. 3) for *L. seeligeri*, *L. ivanovii* and *L. welshimeri*, the most predominant of the *Listeria* species isolated.

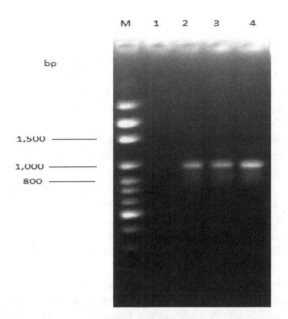

Fig. 3. The PCR products obtained when total genomic DNA from representative of each isolated *Listeria* species were subjected to PCR using siwi2 and Lis1B primer combination for *Listeria seeligeri*, *L. ivanovii* and *L. welshimeri*. A PCR product of 1,200bp was observed. Lanes: m, molecular weight standard; 1, control reaction (all reagent ingredients except chromosomal DNA); 2, *L. ivanovii*; 3, *L.welshimeri*; 4, *L.seeligeri* . PCR products were separated in a 1.5% agarose gel and stained with ethidium bromide.

4. CONCLUSION

This study has demonstrated the presence and distribution of *L. monocytogenes* and other *Listeria* species in raw meat examined in Port Harcourt. The high percentage of the isolation of *Listeria* suggests the need for improved food safety through proper hygienic measures during processing of food and meat products.

ACKNOWLEDGEMENTS

The authors are grateful to the Polish Collection of Microorganisms, Poland for providing the *L. monocytogenes* PCM 2191 and the Nigerian Institute of Medical Research (Molecular Biology and Biotechnology Division), Lagos-Nigeria for the use of their laboratory facilities and technical assistance.

COMPETING INTERESTS

Authors have declared that no competing interests exists

REFERENCES

1. Adzitey F, Huda N. *Listeria monocytogenes* in foods: incidence and possible control measures. Afr. J. Microbiol. Res. 2010;4(25):2848-2855.
2. Barbuddhe SB, Malik SVS, Bhilegaonkar KN, Kumar P, Gupta LK. Isolation of *Listeria monocytogenes* and anti-listeriolysin o detection in sheep and goats. Small Rum. Res. 2000;38(2):151-155.
3. Border PM, Howard JJ, Plastow GS, Siggens KW. Detection of *Listeria* species and *Listeria monocytogenes* using polymerase chain reaction. Lett. Appl. Microbiol. 1990;11(3):158-162.
4. Bubert A, Hein I, Raugh M, Lehner A,Yoon B, Goebel W, Wagner M. Detection and differentiation of *Listeria* spp. by a single reaction based on multiplex PCR. Appl. Environ. Microbiol. 1999;65(10):4688-4692.
5. Filiousis G, Johansson A, Frey J, Perreten V. Prevalence, genetic diversity and antimicrobial susceptibility of *Listeria monocytogenes* isolated from open-air food markets in Greece. Food Control. 2009;20(3):314-317.
6. Hassan EL-S, Ferage M, Nahla T. Lactic acid and ph as indication for bacterial spoilage of meat and some meat products. J Appl. Sci.Res. 2006;2(8):522-528.
7. Hitchins AD, Jinneman KC, Yoshitomi KJ, Blackstone GMK, Thammasouk K, Johnson JM, et al. Multiplex real–time PCR to simultaneously detect Listeria spp. and *L. monocytogenes* from a variety of food enrichments, Abstr. 49. XV International Symposium on problems of listeriosis (ISOPOL), Uppsala; 2004.
8. Ikeh MAC, Obi SKC, Ezeasor DN, Ezeonu IM, Moneke AN. Incidence and pathogenici-ty profile of L*isteria* sp. isolated from food and environmental samples in Nsukka, Nigeria. Afr. J.Biotech. 2010;9(30):4776-4782.
9. Indrawattana N, Nibaddhasobon T, Sookrung N, Chongsa-Nguan M, Tungtrongchitr A, Makino S, et al. Prevalence of *Listeria monocytogenes* in raw meats marketed in bangkok and characterization of the isolates by phenotypic and molecular methods. J. Health Popul. Nutr. 2011;29(1):26-38.
10. Ivanović S, Radanović O, Pavlović I, Žutić J. Goats –source of Liste*ria monocytogenes*. Proceedings, IV International Symposium of livestock production, Struga, Makedonija; 2009.
11. Jay JM. Prevalence of *Listeria* spp. in meat and poultry products. Food ControL. 1996;7(4/5):209-214.
12. Johnson JL, Doyle MP, Cassens RG. *Listeria monocytogenes* and other *Listeria* spp. in meat and meat products. J. Food Protect. 1990;53(1):81-91.
13. Kamat AS, Nair PM. Incidence of *Listeria* species in Indian seafoods and meat. J. Food Safety. 1993;14(2):117-130.
14. Karakolev R. Incidence of *Listeria monocytogenes* in beef, pork, raw-dried and raw-smoked sausages in BULGARIA. Food Control. 2009;20(10):953-955.
15. Kim JW, Rajagopal SN. Antimicrobial activities of *Lactobacillus crispatus* ATCC 33820 and *Lactobacillus gasseri* ATCC 33323. The J.Microbiol. 2001;39(2):146-148.
16. Kinnean PR, GRAY CD. SPSS for windows made simple. phychology press, Hove, UE; 1999.
17. Kosek- Paszkowska K, Bania J, Bystron J, Molenda J, Czerw M. Occurrence of *Listeria* sp. in raw poultry meat and poultry meat products. Bull. Vet. Inst. Pulawy. 2005;49:219-222.

18. Laboratory Quality Assurance Division (LQAD). Isolation and identification of *Listeria monocytogenes* from red meat, poultry, egg, and environmental samples. MLG 8.06; 2008.

19. Liu D, Lawrence LM, Austin FW, Ainsworth AJ. A multiplex pcr for species- and virulence-specific determination of *Listeria monocytogenes*. J. Microbiol. Methods. 2007;71(2):133-140.

20. Low JC, Donachie W. A review of *Listeria monocytogenes* and listeriosis. The Vet J. 1997;153(1):9-29.

21. Mena C, Almeida G, Carneiro L, Teixeira P, Hogg T, Gibbs PA. Incidence of *Listeria monocytogenes* in different food products commercialized in Portugal. Food Microbiol. 2004;21(2):213-216.

22. Molla B, Yilma R, Alemayehu D. *Listeria monocytogenes* and other *Listeria* species in retail meat and milk products in Addis Ababa, Ethiopia. Ethiop. J. Health Dev. 2004;18(3):208-212.

23. Nightingale KK, Schukken YH, Nightingale CR, Fortes ED, Ho AJ, Her Z, et al. Ecology and transmission of *Listeris monocytogenes* infecting ruminants and in the farm environment. Appl. Environ. Microbiol. 2004;70(8):4458-4467.

24. Nørrung B, Andersen JK, Schlundt J. Incidence and control of *Listeria monocytogenes* in foods in Denmark. Int. J. Food Microbiol. 1999;53(2-3):195–203.

25. Okutani A, Okada Y, Yamamoto S, Igimi S. Overview of *Listeria monocytogenes* contamination in Japan. Int. J. Food Microbiol. 2004;93(2):131-140.

26. Pagotto F, Daley E, Farber J, Warburton D. Isolation of *Listeria monocytogenes* from all food and environmental samples. Government of Canada. MFHPB-30; 2001.

27. Pesavento G, Ducci B, Nieri D, Comodo N, Lo Nostro A. Prevalence and antibiotic sus-ceptibility of *Listeria* spp. isolated from raw meat and retail foods. Food Control. 2010;21(5):708–713.

28. Rahimi E, Yazdi F, Farzinezhadizadeh H. Prevalence and antimicrobial resistance of *Listeria* species isolated from different types of raw meat in Iran. J. Food Protect. 2012;75(12):2223-2227.

29. Ramaswamy V, Cresence VM, Rejitha JS, Lekshmi MU, Dharsana KS, Prasad SP, et al. *Listeria*-review of epidemiology and pathogenesis. J. Microbiol. Immunol. Infect. 2007;40(1):4-13.

30. Seeliger HPR, Jones D. *Listeria*. In Bergey's Manual of Systematic Bacteriology, Edited by P. H. A. Sneath, N. S. Nair, N. E. Sharpe & J. G. Holt. Balti-more: Williams and Wilkins. 1986;2:1235–1245.

31. Swaminathan B, Gerner-Smidt P. The epidemiology of human listeriosis. Microbes Infect. 2007;9(10):1236-1243.

32. Uyttendaele M, De Troy P, Debevere J. Incidence of *Listeria monocytogenes* in different types of meat products on the Belgian retail market. Int. J. Food Microbiol. 1999;53(1):75-80.

33. Vazquez-Boland JA, Kuhn M, Berche P, Chakraborty T, Dominguez-Bernal G, Goebel W, et al. *Listeria* pathogenesis and molecular virulence determinants. Clin. Microbiol. Rev. 2001;14(3):584–640.

34. Vitas AI, Aguado V, Garcia-Jalon I. Occurrence of *Listeria monocytogenes* in fresh and processed foods in Navarra (Spain). Int. J. Food Microbiol. 2004;90(3):349–356.

35. Wiedmann M, Mobini S, Cole JR, Watson CK, Jeffers G, Boor KJ. Molecular investigation of a listeriosis outbreak in goats caused by an unusual strain of *Listeria monocytogenes*. J. Am. Vet. Med. Assoc. 1999;215(3):369–371.

36. Yucel N, Citak S, Onder M. Prevalence and antibiotic resistance of *Listeria* species in meat products in Ankara, Turkey. Food Microbiol. 2005;22(2-3):241–245.

Screening of Filamentous Fungi for Xylanases and Cellulases Not Inhibited by Xylose and Glucose

L. F. C. Ribeiro[1], L. F. Ribeiro[1], J. A. Jorge[2] and M. L. T. M. Polizeli[2*]

[1]Immunology and Biochemistry Department of Faculdade de Medicina de Ribeirão Preto – USP, Ribeirão Preto, SP, Brazil.
[2]Biology Department of Faculdade de Filosofia Ciências e Letras de Ribeirão Preto – USP, Ribeirão Preto, SP, Brazil.

Authors' contributions

This work was carried out in collaboration between all authors. Author LFCR participated in all operations of this manuscript. This work is part of her Doctor thesis. Author LFR carried out the Lichenase and laccase assays. He performed the analysis of data concerning these enzymes. Authors JAJ and MLTMP designed the study and wrote the protocol performed. Author MLTMP revised the manuscript and she is final responsible for all information presented. All authors read and approved the final manuscript.

ABSTRACT

Aims: Screening different filamentous fungi for thermostable xylanases and cellulases that would not be inhibited by xylose and glucose, respectively.

Methodology: Samples of fungi collected in the Atlantic forest region, Minas Gerais, Brazil, and some fungi from our Culture Collection were used in this screening. All fungi were grown in liquid media containing 1% sugar cane bagasse (SCB). After that, an aliquot of the crude broth was incubated at different temperatures (from 4 to 60 °C) in carboxymethyl cellulose (CMC) or xylan-media plates, for 12 hours. After this period, the plates were stained with Congo Red. Fungi that presented the best results (larger halos) were tested for the effect of adding xylose and glucose in the xylanase and cellulases activities, respectively. Crude extracts obtained from fungi grown in SCB were used for laccase and lichenase assay.

Results: The screening on agar plates with CMC/xylan presented halos of different sizes. From all tested fungi, the best cellulase producer was *Malbranchea pulchella,* which also

*Corresponding author: Email: polizeli@ffclrp.usp.br

presented the most thermostable xylanase. *Penicillium griseofulvum* presented bigger halos at all temperatures tested, but the xylanase lost almost 14% of its stability in higher temperatures. The effect of xylose and glucose on the enzymatic activities recorded dose-dependent. It was observed that 20% activation of the enzymes produced by *M. pulchella* with 30 mM glucose or 20 mM xylose to cellulase and xylanase, respectively. It was observed a loss of less than 20% for *P. griseofulvum* xylolytic activity using 50 mM xylose. Lichenase was detected in some fungi prospected but laccase was not detected.
Conclusion: *Malbranchea pulchella* was a good producer of xylanase and cellulase tolerant to xylose and glucose, respectively. Other studies must be performed with this fungus so that it can be used in the future for biotechnological purposes.

Keywords: Malbranchea pulchella; xylanase; cellulases; xylose inhibition; glucose inhibition.

1. INTRODUCTION

Microbial enzymes have been commercially exploited and successfully used on industrial scale to catalyze several chemical processes [1]. In this context, filamentous fungi that are good enzyme producers are particularly interesting due to their easy cultivation and high production of extracellular enzymes with large industrial potential [2]. These enzymes may be applied in detergent, drinks and food, textile, animal feed, barking, pulp and paper, chemical and biomedical product industries [3,4].

Plant cell wall is a complex structure that surrounds and protects the cell. Its major structural components are cellulose, hemicelluloses and lignin. The degradation of this complex structure requires a complete enzymatic system that includes cellulases, xylanases ligninases and laccase [5]. Cellulolytic enzymes, which hydrolyze cellulose releasing glucose, can be divided into three categories: endoglucanase (endo-1, 4-β-D-glucanase, EG, EC 3.2.1.4); cellobiohydrolase or exoglucanase (exo-1, 4-β-D-glucanase, CBH, EC 3.2.1.91) and β-glucosidase (1, 4-β-D-glucosidase, BG, EC 3.2.1.21) [6,7]. Lichenase, a cellulolytic enzyme, endo-1,3-1,4-β-glucanase, hydrolyzes the internal 1,4-β-glucosyl linkages when the glucosyl residue is linked at the O-3 position [8]. Xylanases degrade plant fibers made of xylan hemicellulose releasing xylose monomers or oligomers [3]. They comprise an enzymatic complex composed by endo-β-1,4-xylanase (1,4-β-D-xylan xylanohydrolase, E.C. 3.2.1.8), β-D-xylosidase (1,4-β-xylan xylanohydrolase, EC 3.2.1.37), and acting together with debranching enzymes (esterases) [2]. Laccase (p-diphenol:oxygen oxidoreductase, EC 1.10.3.2) is a copper-containing oxidase that catalyzes the reduction of molecular oxygen to water, bypassing a stage of hydrogen peroxide production [9].

Cellulases and xylanases have great potential for industrial application, like bioconversion of lignocelluloses into fermentable sugar that may be used by yeasts to produce ethanol [10,11]. Laccase is an interesting enzyme in this context because it may fragment the lignin, releasing cellulose and hemicelluloses, easing the action of cellulases and xylanases [12].

Brazil has a large diversity of microorganisms due to the fact it comprises a vast physical territory where there are large areas of forests and vegetal diversity. There is a possibility of the existence of some fungi that are excellent producers of cellulases, xylanases and laccases that remain unknown. The aim of this work was to select filamentous fungi from nature in the region of Uberlândia-MG and from the mycoteca of our laboratory that are good producers of xylanases, cellulases and laccases that are thermostable and not inhibited by their products.

2. MATERIALS AND METHODS

2.1 Microorganism

Filamentous fungi were isolated from decomposing trees in Atlantic forest region, Minas Gerais, Brazil. Fungal identification was performed at Departamento de Micologia da Universidade Federal de Pernambuco, Brazil. Stock cultures were maintained at 4°C in oat extract agar media and constant replications were performed at regular intervals

2.2 Screening on Agar Medium

Filamentous fungi were cultivated for 72 hours at optimum temperature of each fungus in modified SR [13] liquid medium. Ten milligrams yeast extract and 10 mg of peptone were added for 100 mL medium. The medium was modified to make a not so rich medium, so the fungus could grow using mainly the sugarcane bagasse as the carbon source and not yeast extract. Media were filtrated and used as enzymatic extract. A volume of 10 μL of extracts was placed inside a cavity of 7 mm diameter made in an agar medium. This medium contained the same salt solution used for SR medium added with 0.5% (w/v) xylan Birchwood (SIGMA) or 0.5% (w/v) carboxymethylcellulose sodium salt (CMC) with low viscosity (SIGMA). The cavity was filled on its base with agar 1% (w/v). These plates (90 mm x 15 mm) containing about 0.5 cm of agar layer depth were incubated at 30, 40 and 50 °C for 15 hours and stained with Congo Red [14].

2.3 Xylose and Glucose Effect on Xylanase and Cellulose Activities

Enzymatic assays were done by adding xylose or glucose in different concentrations (0 to 70 mM) in the reaction mixture. The substrate for xylanase was 0.2 % Remazol Brilliant Blue R-D-Xylan and for cellulose was 0.2 % Remazol Brilliant Blue Carboxymethylcellulose [15]. Both reactions were carried out with 100 mM sodium acetate buffer pH 5.0 and incubated at 50°C for 5 and 30 minutes, respectively. The reaction was stopped with the addition of 2 volumes of ethanol absolute. After centrifugation at 4000 x g for 5 minutes, the absorbance of the supernatant was read at 595 nm.

2.4 Quantitative Assay for Xylanase Activity

The amount of xylanase produced was measured by using 1% xylan Birchwood, SIGMA, as substrate [16]. Xylanase activity was assayed in 200 μL of a reaction mixture containing 50 μL of crude broth, 100 μL of 1% xylan from Birchwood (prepared in 0.1 M sodium acetate buffer, pH 5.0) and 50 μL of 0.1 M sodium acetate buffer, pH 5.0. The mixture was incubated at 50°C for 5 min. The reaction was stopped by the addition of 200 μL of 3, 5-dinitrosalicylic acid (DNS) and the contents were boiled for 5 min [17]. The absorbance at 540 nm was read and the amount of reducing sugars released was quantified using xylose as standard. One unit of enzyme activity is defined as the amount of enzyme that releases 1 μmol of xylose in 1 min under the assay conditions.

2.5 Quantitative Assay for Cellulase Activity

Cellulase (CMCase) activity was determined by mixing 100 µL of 1% (w/v) Carboxymethyl cellulose sodium salt, medium viscosity, SIGMA with 0.7 degree of substitution, (prepared in 0.1 M sodium acetate buffer pH 5.0) with 150 µL of crude broth and 50 µL 0.1M sodium acetate buffer. The reaction mixture was incubated at 50°C for 30 min. The reaction was stopped by the addition of 300 µL of 3, 5-dinitrosalicylic acid (DNS) and the contents were boiled for 5 min. The absorbance at 540 nm was read and the amount of reducing sugars released was quantified using glucose as standard. One unit of enzyme activity is defined as the amount of enzyme that releases 1 µmol of glucose in 1 min under the assay conditions.

2.6 Quantitative Assay for Lichenase Activity

β-1,3–1,4-Glucanase activity was assayed by the determination of reducing sugars released from lichenan substrate (MP Biomedical – Solon, USA) using the 3,5-dinitrosalicylic acid (DNS) method [17]. The assay mixture (0.5% (w/v) lichenan, 50 mM MES - 2-(N-morpholino) ethanesulfonic buffer, pH 6.0) was incubated with the crude broth for 10 min and the reaction was stopped by the addition of DNS reagent. The absorbance at 540 nm was read and the amount of reducing sugars released was quantified using glucose as standard. One unit of enzyme activity is defined as the amount of enzyme that releases 1 µmol of glucose in 1 min under the assay conditions.

2.7 Quantitative Assay for Laccase Activity

Laccase activity was determined by the rate of oxidation of 2,2'-azinobis-(3-ethylbenzthiazoline-6-sulfonic acid) (ABTS), which was monitored at 420 nm (ε = 36000 M^{-1}cm^{-1}) [18]. The assay mixture contained crude broth, 1 mM ABTS and 50 mM acetate buffer, pH 4.5.

2.8 Soluble Protein Assay

Protein content of the culture supernatant was determined according to the method described by Bradford [19] using bovine serum albumin (BSA) as standard.

2.9 Analysis of Reaction Products by TLC – Thin Layer Chromatography

Xylan and CMC degradation products were qualitatively determined by thin-layer chromatography (TLC) on precoated TLC sheets (silica gel; DC-Alufolien Kieselgel 60, Merck) revealed with n-butanol-ethanol-water (5:3:2, vol/vol/vol). The products were visualized by spraying the layers with a 1:1 (vol/vol) mixture of 0.2% methanolic orcinol and 20% sulfuric acid [20].

3. RESULTS AND DISCUSSION

The screening on agar plates with CMC or xylan media presented many halos with different sizes (Table 1). This experiment was done to select the fungus that produced enzymes with the best activities and higher thermostability. From all the tested fungi, the best cellulolytic enzyme producers were *Malbranchea pulchella* and *Penicillium griseofulvum*. Among those, *M. pulchella* presented the higher stability at 50°C (Fig. 1 and Table 1), which can be indirectly measured by the size of the halo. Since this fungus is classified as a thermophilic mold [21], this result was expected. The best producer of xylolytic enzymes was *Aspergillus clavatus*, that presented bigger halos at all the temperatures tested, but it lost almost 14% of its stability in higher temperatures (50°C).

Table 1. Diameter of halos produced by the hydrolysis of CMC or xylan from Birchwood

Specie	Halo diameter in CMC (cm)				Halo diameter in xylan (cm)			
	4°C	30°C	40°C	50°C	4°C	30°C	40°C	50°C
M. pulchella	0 ±0	1.30 ±0.1	1.63±0.1	1.82 ±0.1	0 ±0	1.80 ±0.1	1.94 ±0.0	2.22 ±0.1
T. longibrachiatum	0 ±0	1.22 ±0.1	1.41±0.0	1.01 ±0.0	0 ±0	1.82 ±0.0	2.05 ±0.1	2.03 ±0.1
A. clavatus	0 ±0	0 ±0	1.13±0.1	1.34 ±0.1	1.21±0.0	2.89 ±0.1	2.61 ±0.0	2.51 ±0.0
A. terreus albino	0 ±0	1.30 ±0.0	1.50±0.0	1.61 ±0.0	0 ±0	1.47 ±0.0	1.73 ±0.0	1.72 ±0.0
F. oxysporum	0 ±0	1.2 ±0.1	1.42±0.1	1.40 ±0.1	0 ±0	1.01 ±0.1	1.28 ±0.1	0 ±0
P. griseofulvum	1.0±0.0	1.61 ±0.1	1.78±0.1	1.61 ±0.1	0 ±0	1.62 ±0.0	1.83 ±0.0	1.60 ±0.0
Aspergillus sp	0 ±0	1.22 ±0.1	1.39±0.0	1.0 ±0.1	0 ±0	1.31 ±0.0	1.51 ±0.1	1.52 ±0.1

The experiment was conducted in triplicate.

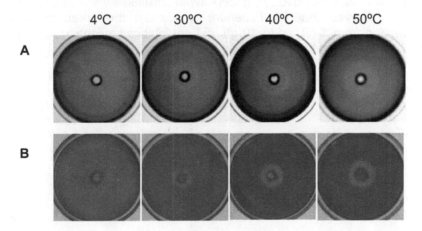

Fig. 1. *M. pulchella* halos in xylan (A) and CMC (B) after the incubation at different temperatures. Plates were incubated at the temperatures for 15 hours

The halo detection generated by the enzymatic hydrolysis is an easy and valuable way to do this screening. There is still a resistance in using methods of screening for filamentous fungi. Some authors use this technique by inoculating the fungus directly on the plate and measuring the generated halo [22,23], but this assay is not always straight forward because depending on the fungus, it may sporulate too much and the spores may spread over the plate as previously seen in our results (not shown). If this happens, more than a unique colony will grow on the plate, making it confusing to interpret the results.

Another advantage of our screening method is that it was possible to qualitatively analyze the thermostability by incubating the plates at different temperatures. It saves time and work because for the usual screening method it is necessary to incubate a certain amount of crude broth at different temperatures and only after that the enzymatic activity is measured [24]. When the value of cellulase activity is low, it will be even more difficult to observe differences concerning the thermostability.

3.1 Glucose and Xylose Can Enhance Glycosyl Hydrolase Activities

Another parameter that was interesting to investigate with *M. pulchella* and *P. griseofulvum* was the effect of monosaccharides, the end products of the action of the xylolytic and cellulolytic complexes. So, those activities were measured with different concentrations of xylose and glucose, respectively. As *M. pulchella* was stable at the highest temperature tested (50 °C, Table 1) and *P. griseofulvum* presented good results for xylanase and cellulose activities (Table 1), they were selected to test the effect of monosaccharides amendment to the reaction mixture. *A. clavatus* was discarded despite having the highest activity in xylan but it showed low activity in CMC (Table 1).

The results obtained showed that it seems to exist an activation of the enzymes produced by *M. pulchella* (Fig. 2, A). This activation was dose dependent, but on the other hand the inhibition of these enzymes by their end products (Figs. 2, A and B) was not observed. For the xylolytic activity of *P. griseofulvum* and for both enzymatic activities (xylolytic and cellulolytic) of *M. pulchella*, it was observed a loss of less than 20% from 20 mM and 40 mM of the monosaccharides concentration, respectively (Figs. 2, A and B) and from this on their activities remained constant. However, *P. griseofulvum* cellulases were inhibited by glucose. Xiros et al. [25] observed that for *Fusarium oxysporum* the presence of xylose in concentrations varying from 1 g/L up to 10 g/L did not inhibited xylanase activity, as it has been seen for *P. griseofulvum* in this work (Fig. 2 B).

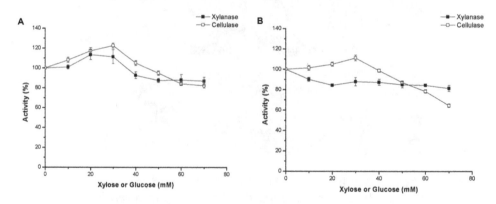

Fig. 2. Xylose or glucose effect on xylolytic and cellulolytic activities, respectively. This assay was carried out with Remazol Brilliante Blue Xylan or CMC. A – *Malbranchea pulchella*; B – *Penicillium griseofulvum*

3.2 Elucidating the End Products of Xylanases and Cellulases

In order to investigate the hydrolysis products released by the enzymes and to better understand the effect of monosaccharides on xylolytic and cellulolytic activities, an enzyme assay using xylan or CMC as substrates was done. After a period of up to 12 hours of incubation at 50°C, the crude broths were examined by TLC (Fig. 3). The results revealed that for *M. pulchella*, xylose was not released even after 12 hours of incubation. However, glucose was detected with incubation of 12 hours, but not before this time. This may be the reason that there were not strong negative effects of those sugars over xylanases and cellulases activities (Fig. 2). Those enzymes are usually inhibited by their direct products [26], but it was not observed in this work. In order to *Penicillium griseofulvum*, it was not possible to observe the presence of both monosaccharides even after 12 hours of incubation (Fig. 3). Also, the effect of these sugars over the enzymatic activity did not present a high inhibition (Fig. 2 B), moreover cellulolytic activity showed a discrete activation with 30 mM glucose (Fig. 2 B).

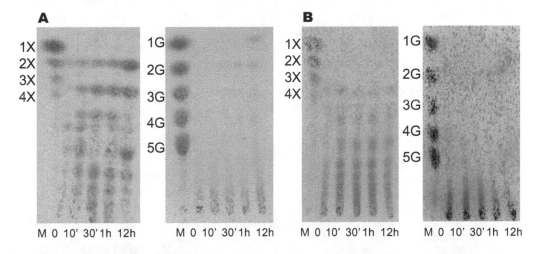

Fig. 3. TLC of xyolytic and cellulolytic enzymes, respectively. A – *Malbranchea pulchella*; B - *Penicillium griseofulvum*. 1 to 4 X represent the monomer xylose and the oligosaccharides containing 2 to 4 residues of xylose. The same for 1 to 5 G, except that it is for glucose. M – pattern of the mono and some oligosaccharides. 0 to 12h – time of incubation of the crude broth with the substrate (xylan or CMC)

3.3 Investigation of Lichenase and Laccase Activities

All seven fungi (Table 1) were grown in liquid medium containing sugar cane bagasse as carbon source because it is a complex source that could allow the screening of xylanases, cellulases and maybe ligninases. Therefore, screening for lichenase and laccase was also carried out (Table 2). Results of the xylanase, cellulase, lichenase and laccase activities are shown in Table 2. *A. clavatus* presented the highest lichenase activity (an endoglucanase that hydrolyzes specifically β1-4 linkages followed by β1-3 bonds), followed by *A. terreus* albino, *F. oxysporum* and *P. griseofulvum,* that presented lichenase activity higher 3 U/mL. These results are very useful because there is a scarcity of data in the literature regarding filamentous fungi that secrete lichenases. Laccase was not detected in any fungi, although it is mainly produced by filamentous fungi [27]. This enzyme was assayed in an attempt to make a relation between hemicellulosic activities with the degradation of lignin. In this way, the aim of this experiment was to check if a fungus that possesses laccase activity would be a better producer for xylolytic and cellulolytic enzymes. Unfortunately it was not possible to do this kind of comparison since laccase activity was not detected.

As it can be seen in the Table 2, all the fungi analyzed presented high values of xylanase activity when grown in sugarcane bagasse as the carbon source (Table 2). Souza et al. [28] found much reduced xylanase activity (smaller than 1 U/mL) when they grew *A. niveus* in sugarcane bagasse as carbon source [28], however, Mahamud and Gomes recorded it over 9 U/mL for *Trichoderma* sp grown with 2% sugarcane bagasse [29]. In order to *A. niger* it was about 3 U/mL, smaller than the values obtained in this work in which the xylanase activity varied from 6.81 to 19.93 U/mL (Table 2).

The results for CMCase activity were very poor in comparison to the earlier reports in literature. The activity obtained in this work were lower than 0.2 U/mL (Table 2), while the literature shows activities varying from close to 0.25 U/mL [28] for *A. niger* until less than 0.3 U/mL for *Trichoderma viride* when the fungi were grown in sugarcane bagasse [30]. So it seems that the fungi tested are not good producers of cellulolytic enzymes when grown in sugarcane bagasse.

Table 2. Screening of lignocellulolytic enzymes for fungi grown in liquid medium containing sugar cane bagasse as carbon source

Sample	Specie	Enzyme Activity (U/mL)			
		Cellulase	Xylanase	Lichenase	Laccase
M. pulchella	*Malbranchea pulchella*	0.049 ±0.008	9.2 ±0.3	1.2 ±0.09	ND
LF212	*Trichoderma longibrachiatum*	ND	11.8 ±0.8	2.0 ±0.2	ND
Acla	*Aspergillus clavatus*	ND	19.93 ±1.28	5.45 ±0.41	ND
Ateralb	*Aspergillus terreus albino*	0.162 ±0.025	10.18 ±0.71	3.47 ±0.23	ND
FusRosa	*Fusarium oxysporum*	0.070 ±0.011	11.78 ±0.58	3.64 ±0.27	ND
V4	*Penicillium griseofulvum*	0.032 ±0.009	6.81 ±0.16	4.44 ±0.27	ND
LF32	*Aspergillus* sp.	0.041 ±0.005	9.57 ±0.57	2.15 ±0.11	ND

ND - Not Detected.

4. CONCLUSION

It is important to find good producers of lignocellulolytic enzymes, because of its wide application. These enzymes may be used in paper and pulp industries, bioethanol production, animal feed industry, etc. From all tested fungi, *Malbranchea pulchella* exhibited the best set of results. It presented pronounced halos in CMC and in xylan, as well as high thermostability and it was not inhibited by glucose or xylose. Actually, it presented a little activation in low concentrations of monosaccharides, both for cellulase and xylanase activities, but did not present inhibition by both sugars. This result is interesting because *Malbranchea pulchella* can be used in biorefineries aiming the production of second generation ethanol.

ACKNOWLEDGEMENTS

This work was supported by grants from Fundação de Amparo à Pesquisa do Estado de São Paulo (FAPESP), Conselho de Desenvolvimento Científico e Tecnológico (CNPq) and the National System for Research on Biodiversity (Sisbiota-Brazil, CNPq 563260/2010-6/FAPESP n° 2010/52322-3). J.A.J. and M.L.T.M.P. are Research Fellows of CNPq. L.F.C.R. and L.F.R. are recipients of FAPESP Fellowships. We thank Ricardo Alarcon and Mauricio de Oliveira for technical assistance and Mariana Cereia for English support.

COMPETING INTERESTS

Authors have declared that no competing interests exist.

REFERENCES

1. Sohail M, Naseeb S, Serwani SK, Sultana S, Aftab S, Shahzad S, Ahmad A, Khan SA. Distribution of hydrolytic enzymes among native fungi: *Aspergillus* the pre-dominant genus of hydrolase producer. Pak. J. Bot. 2009;41(5):2567-2582.
2. Polizeli ML, Rizzatti AC, Monti R, Terenzi HF, Jorge JA, Amorim DS. Xylanases from fungi: properties and industrial applications. Appl. Microbiol. Biotechnol. 2005;67:577-591.
3. Guimarães LHS, Peixoto-Nogueira SC, Michelin M, Rizzatti ACS, Sandrim VC, Zanoelo F, Aquino ACMM, Junior AB, Polizeli MLTM. Screening of filamentous fungi for production of enzymes of biotechnological interest. Braz. J. Microbiol. 2006; 37:474-480.
4. Whiteley CG, Lee DJ. Enzyme technology and biological remediation. Enz. Microb. Technol. 2006;38:291-316.
5. Caffall KH, Pattathil S, Phillips SE, Hahn MG, Mohnen D. *Arabidopsis thaliana* T-DNA mutants implicate GAUT genes in the biosynthesis of pectin and xylan in cell walls and seed testa. Mol. Plant. 2009;2,1000–1014.
6. Li YH, Ding M, Wang J, Xu GJ, Zhao F. A novel thermoacidophilic endoglucanase, Ba-EGA, from a new cellulose degrading bacterium, *Bacillus* sp. AC-1. Appl. Microbiol. Biotechnol. 2006;70:430-436.
7. Gao J, Weng H, Zhu D, Yuan M, Guan F, Xi Y. Production and characterization of cellulolytic enzymes from the thermoacidophilic fungal *Aspergillus terreus* M11 under solid-state cultivation of corn stover. Bioresour. Technol. 2008;99:7623-7629.
8. Pang Z, Kang YN, Ban M, Oda M, Kobayashi R, Ohnishi M, Mikami, B. Crystallization and preliminary crystallographic analysis of endo-1,3-β-glucanase from *Arthrobacter* sp. Acta Crystallogr Sect F Struct Biol Cryst Commun. 2005;61:68-70.
9. Morozova OV, Shumakovich GP, Gorbacheva MA, Shleev SV, Yaropolov AI. "Blue" laccases. Biochemistry. 2007;72,1136-1150.
10. Viikari L, Kantelinen A, Sundqvist J, Linko M. Xylanases in bleaching: from an idea to the industry. FEMS Microbiol Rev. 2001; 13: 335-350.
11. Gomes J, Purkarthiofer HM, Kapplomiller J, Sinnner M, Steiner W. Production of high level of cellulase–free xylanase by the thermophilic fungus, *Thermomyces lanuginosus* in laboratory and pilot scales using lignocellulosic materials. Appl. Microbiol. Biotechnol. 1993;39:700-707.
12. Couto SR, Herrera JLT. Industrial and biotechnological applications of laccases: A review. Biotechnol. Adv. 2006;24:500-513.
13. Rizzatti ACS, Sandrim VC, Jorge JA, Terenzi HF, Polizeli MLTM. Influence of temperature on the properties of the xilanolytic enzymes of the thermotolerant fungus *Aspergillus phoenicis*. J. Ind. Microbiol. Biotechnol. 2004;31:88-93.
14. Teather RM, Wood PJ. Use of congo red-polysaccharide interactions in enumeration and characterization of cellulolytic bacteria from the bovine rumen. Appl. Environ. Microbiol. 1982;4:777-780.
15. Biely P, Mislovicová D, Toman R. Soluble chromogenic substrates for the assay of endo-1,4-beta-xylanases and endo-1,4-beta-glucanases. Anal. Biochem. 1985;144(1):142-146.

16. Bailey MJ, Biely P, Poutanen K. Interlaboratory testing of methods for assay of xylanase activity. J. Biotechnol. 1992;23:257-270.

17. Miller GL. Use of dinitrosalicylic acid reagent for determination of reducing sugar. Anal. Chem. 1959;31(3):426-428.

18. Arnold FH, Georgiou G. Directed enzyme evolution, screening and selection methods. Methods Mol. Biol. 2003;230:3-26.

19. Bradford MM. Rapid and sensitive method for the quantitation of microgram quantities of protein utilizing the principle of protein dye binding. Anal. Biochem. 1976;72:248-254.

20. Fontana JD, Gebara M, Blumel M, Schneider H, Mackenzie CR, Johnson KG. α–4–O–methyl–D–glucuronidase component of xylanolytic complexes. Methods Enzymol. 1988;160,560-571.

21. Banerjee S, Archana A, Satyanarayana T. Xylose metabolism in the thermophilic mould *Malbranchea pulchella* var. *sulfurea* TMD-8. Curr. Microbiol. 1994,29:349-352.

22. Sridevi B, Charya MAS. Isolation, identification and screening of potential cellulose-free xylanase producing fungi. Afr. J.Biotechnol. 2011;10:4624-4630.

23. Abdel-Sater MA, El-Said AHM. Xylan-decomposing fungi and xylanolytic activity in agricultural and industrial wastes. Internat. Biodeter..Biodegrad. 2001;44:15-21.

24. Alves-Prado HF, Pavezzi FC, Leite RS, de Oliveira VM, Sette LD, Dasilva R. Screening and production study of microbial xylanase producers from Brazilian Cerrado. Appl Biochem Biotechnol. 2010;161:333-346.

25. Xiros C, Katapodis P, Chirstakopoulos P. Factors affecting cellulose and hemicelluloses hydrolysis of alkali treated brewers spent gain by *Fusarium oxysporum* enzyme extract. Bior. Technol. 2011;102:1688-1696.

26. Andrić P, Meyer AS, Jensen PA, Dam-Johansen K. Effect and modeling of glucose inhibition and in situ glucose removal during enzymatic hydrolysis of pretreated wheat straw. Appl. Biochem. Biotechnol. 2010;160:280-297.

27. Xu F, In Flickinger MC, Brew SW. Encyclopedia of bioprocessing technology: fermentation, biocatalysis and bioseparations. Wiley, New York. 1999;1545-1554.

28. Souza WR, Gouveia PF, Savoldi M, Malavazi I, Bernardes LAS, Goldman LH, de Vries RP, Oliveira JVC, Goldman GH. Transcriptome analysis of *Aspergillus niger* grown on sugarcane bagasse. Biotechnol. Biofuels. 2011;4:40.

29. Mahamud MR, Gomes DJ. Enzymatic saccharification of sugar cane bagasse by the crude enzyme from indigenous fungi. J. Sci. Res. 2011;4(1),227-238.

30. Ahmed FM, Rahman SR, Gomes DJ. Saccharification of sugarcane bagasse by enzymatic treatment for bioethanol production. Malaysian J. Microbiol. 2012;8(2):97-103.

Changes in Microbial Population of Palm Oil Mill Effluent Polluted Soil Amended with Chicken Droppings and Cow Dung

L. O. Okwute[1*] and U. J. J. Ijah[2]

[1]Department of Biological Sciences, University of Abuja, P. M. B. 117, Gwagwalada-Abuja, Nigeria.
[2]Department of Microbiology, Federal University of Technology, P. M. B. 65, Minna-Niger State, Nigeria.

Authors' contributions

This work was carried out in collaboration between both authors. Both authors read and approved the final manuscript.

ABSTRACT

Aim of Study: To assess changes in microbial population in palm oil mill effluent (POME) polluted soil amended with chicken droppings and cow dung.
Study Design: 32 plots measuring 4 m^2 were mapped out in a randomized complete block design of five main plots with three replicates. Data collected were subjected to ANOVA using SPSS.
Place and Duration of Study: Faculty of Agriculture, Kogi State University, Anyigba, Kogi State, Nigeria: July 2011 to November 2011.
Methodology: Plots were polluted with palm oil mill effluent and subsequently remedied using varying amounts of chicken droppings and cow dung (5 kg, 10 kg and 15 kg). Microbiological analysis was carried out using Nutrient agar and Sabouraud dextrose agar for the enumeration of total aerobic heterotrophic bacteria (TAHB) and fungi (moulds and yeasts) respectively.
Results: Significant difference ($P=0.05$) in TAHB counts after 1 month and 2 months in all treatments with the exception of unamended polluted and unpolluted control soils. The counts however, increased after 2 months in all treatments with the exception of unamended polluted soil. The overall data suggest that amendment of the POME polluted soil enhanced microbial growth, particularly after 2 months meaning that bioremediation of

---

*Corresponding author: Email: lolookwute@yahoo.com

the polluted soil can be achieved with the organic wastes within a short time.

Conclusion: Chicken droppings (at 10 kg and 15 kg/4m^2 plot) and a combination of chicken droppings and cow dung (at 10 kg and 15 kg/4m^2 plot) have the ability to significantly increase microbial populations in palm oil mill effluent (POME) polluted soil thereby stimulating the bioremediation of the polluted soil.

Keywords: Palm oil mill effluent (POME); chicken droppings; cow dung.

1. INTRODUCTION

Palm oil processing is carried out in mills where oil is extracted from the palm fruits. Large quantities of water are used during the extraction of crude palm oil from the fresh fruits and about 50% of the water results in palm oil mill effluent (POME) [1]. It is estimated that for 1 tonne of crude palm oil produced, 5 - 7.5 metric tonnes of water will end up as POME [2]. POME is usually discharged into the environment in either a raw or treated state. Raw POME consisting of complex vegetative matter is thick, brownish, colloidal slurry of water, oil and solids including about 2% suspended solids originating mainly from cellulose fruit debris, that is, palm fruit mesocarp [3]. The raw or partially treated POME has an extremely high content of degradable organic matter, which is due in part to the presence of unrecovered palm oil [2]. This highly polluting wastewater can, therefore, cause pollution of waterways due to oxygen depletion and other related effects [2]. It has been reported that heavy application of POME to soil significantly ($P \leq 0.05$) reduced the total aerobic heterotrophic bacterial populations in the soil when compared to counts for non-POME soil samples [4,5]. The POME also reduced phosphate solubilizing, nitrifying and lipolytic bacterial counts [5] and ammonium oxidizers were isolated from non-POME soil samples but not from POME polluted soil samples [6]. Microbial degradation appears to be the most environmentally friendly method of removal of oil pollutants since other methods such as surfactant washing and incineration lead to introduction of more toxic compounds to the environment [7]. The use of organic wastes (chicken droppings and cow dung) as cheap alternatives to procedures such as biopiling, membrane technology and activated sludge reactors is therefore an easy option for local mill operators for the reclamation of arable land. This study aims to assess the changes in microbial population in palm oil mill effluent (POME) polluted soil amended with chicken droppings and cow dung. Microorganisms present in the soil and organic wastes were identified and their potential utilization of POME was determined.

2. MATERIALS AND METHODS

2.1 Collection of Samples

Palm oil mill effluent (POME) was obtained from an established oil mill on the outskirts of Anyigba Town, Kogi State, Nigeria. The effluent which is normally contained in a plastic drum was mixed thoroughly before being transferred into clean plastic containers, tightly screwed and transported to the laboratory in an ice box. When not in use the POME was stored in a refrigerator at 4°C. The organic wastes used were chicken droppings and cow dung. The chicken droppings was collected fresh from a poultry house (deep litter) in Gwagwalada, Abuja, Nigeria while cow dung was collected fresh from Gwagwalada abattoir, Abuja-Nigeria in polythene bags and transported to the laboratory. The organic wastes were sun-dried for 48 hours before being ground and packed in clean polythene bags.

2.2 Study Site

Randomized complete block design, (RCBD) was adopted. The land which was situated in a demarcated and secured area in the Faculty of Agriculture, Kogi State University, Anyigba, Nigeria was flat, non-sloping and well drained. It was ploughed, harrowed and mapped out into 5 main plots (80 m^2, 80 m^2, 80 m^2, 20 m^2, 20 m^2). Three plots (80 m^2 each) representing those for cow dung, chicken droppings and a combination of the two organic wastes were subdivided into 9 sub-plots, each measuring 2 m by 2 m (4 m^2) and a space of free land of 1 m by 2 m on each side of each plot to create adequate gaps (alleys) between plots. The remaining two main plots with an area of 20 m^2 each were subdivided into 3 plots of 2 m by 2 m with a gap (alley) of 1 m by 2 m in between plots. The two plots served as control 1 (soil alone) and control 2 (soil + POME).

2.3 Application of POME (Pollution)

On each sub-plot of 4 m^2, 12 litres of palm oil mill effluent (POME) was applied evenly using a garden watering can. This was done on all plots except control 1 (soil alone) which was left undisturbed. After the POME application, an auger was used to collect soil from all plots into properly labeled, clean polythene bags and transported to the laboratory for analysis. Soil samples were also collected after one month and two months.

2.4 Bioremediation of Polluted Soil

Two weeks after pollution, application of organic wastes was carried out. Cow dung was applied to each subplot measuring 4 m^2 in the following order, 5 kg (3 subplots), 10 kg (3 subplots), 15 kg (3 subplots). This was done by spreading the dried organic wastes evenly on each subplot. The same treatment was given to another set of 9 subplots for chicken droppings in the same order. The remaining 3 subplots received a combination of the two organic wastes in varying proportions (5 kg, 10 kg and 15 kg). No organic waste was applied to two main plots which served as control 1 (soil alone) and control 2 (soil + POME). After application of the wastes, adequate mixing of the wastes with the polluted soil was undertaken. Tilling was repeated once in two weeks throughout the period of the field experiment (two months). Soil samples were collected immediately after the application of the organic wastes and at 1 month interval for a period of two months into properly labeled, clean polythene bags and transported to the laboratory for microbiological analysis.

2.5 Microbiological Analysis

Microorganisms in the soil samples were enumerated by inoculating 0.1 ml serially diluted samples onto nutrient agar (NA) and Sabouraud Dextrose agar (SDA) plates for the enumeration of aerobic heterotrophic bacteria and fungi respectively using the spread plate method. The inoculated NA plates were incubated at 30°C for 48 hours while the SDA plates were incubated at 25°C for 3-5 days. Observed colonies were counted and expressed as colony forming units per gram (cfu/g) of soil.

2.6 Characterization and Identification of Microbial Isolates

2.6.1 Bacterial Isolates

Bacterial isolates were characterized based on Gram reaction and biochemical tests. The biochemical tests included production of coagulase, catalase, indole, urease, motility test, citrate utilization test, starch hydrolysis, Methyl Red-Voges Proskaeur (MR-VP), triple sugar iron test, utilization of sodium azide and various carbohydrates (glucose, lactose, maltose, fructose, mannitol, sucrose and arabinose). The isolates were identified to the species level by comparing their characteristics with those of known taxa, as described in [8].

2.6.2 Mould Isolates

Mould isolates were characterized based on microscopic and macroscopic appearances which comprised pigmentation, colour of aerial and substrate hyphae, type of hyphae, shape and kind of asexual spore, presence of special structures such as foot cell, sporangiophore or conidiophores and the characteristic of the spore head. The identities of the isolates were determined using the scheme of [9].

2.6.3 Yeast Isolates

Yeast isolates were Gram stained and characterized based on colonial morphology, cell micromorphology, germ tube and blastosphore formation, gelatin liquefaction, starch hydrolysis, growth at 37°C and on 50% glucose and fermentation of the following carbohydrates: galactose, glucose, sucrose, maltose and lactose. The identities of the isolates were determined using the scheme of [10].

2.7 Statistical Analysis

Data generated from the study were analyzed using the computer package SPSS (Version 19.0) [11] which employed the use of univariate analysis of variance (ANOVA) at $P = 0.05$ confidence limit.

3. RESULTS AND DISCUSSION

3.1 Bacterial Counts in Soil Samples after Two Months of Bioremediation

The counts were statistically significant ($P = 0.05$) for bacteria in CKD (10 kg and 15 kg) and CD + CKD (10 kg) after 2 months of bioremediation (Table 1). The least bacterial counts was observed in POME polluted soil at zero time of bioremediation but this gradually increased in all treatments as bioremediation proceeded to the second month. The organic wastes with the most significant microbial counts at $P = 0.05$ confidence level were Chicken droppings (10 kg and 15 kg), Cow dung + Chicken droppings (10 kg and 15 kg). It can be seen from the results that chicken droppings generally had an edge in stimulating the growth of microorganisms in the polluted soils. All microbial counts decreased at the time of POME application and gradually increased over the period of bioremediation. This agrees with the findings of [5]. The decrease in counts may be directly related to the acidic nature of raw POME [3]. The gradual increase in microbial counts after the application of the organic wastes indicated that the nutrients in the wastes, possibly nitrogen and phosphorus helped the microorganisms to overcome the initial stress experienced as a result of the POME application.

Table 1. Bacterial Counts in Soil Samples after Two Months of Bioremediation

Treatment	Bacterial counts (cfu/g)		
	Time (Months)		
	0	1	2
A	$3.2 \times 10^5 + 0.002$	$6.0 \times 10^5 + 0.012$	$1.1 \times 10^6 + 0.001$
B	$4.5 \times 10^5 + 0.001$	$9.8 \times 10^5 + 0.012$	$1.4 \times 10^7 + 0.002$
C	$5.4 \times 10^6 + 0.008$	$1.5 \times 10^6 + 0.001$	$1.7 \times 10^7 + 0.014$
D	$1.1 \times 10^6 + 0.05$	$4.5 \times 10^6 + 0.02$	$6.5 \times 10^6 + 0.002$
E	$1.9 \times 10^6 + 0.003$	$7.5 \times 10^6 + 0.003$	$1.9 \times 10^7 + 0.01*$
F	$1.8 \times 10^6 + 0.018$	$2.3 \times 10^7 + 0.01$	$2.6 \times 10^7 + 0.03*$
G	$4.4 \times 10^6 + 0.01$	$1.0 \times 10^6 + 0.001$	$1.0 \times 10^7 + 0.01$
H	$5.5 \times 10^6 + 0.02$	$1.2 \times 10^7 + 0.02$	$1.8 \times 10^7 + 0.002*$
I	$6.7 \times 10^6 + 0.01$	$1.4 \times 10^6 + 0.004$	$1.1 \times 10^7 + 0.02$
J	$2.0 \times 10^4 + 0.01$	$4.0 \times 10^5 + 0.001$	$1.1 \times 10^5 + 0.03$
K	$4.8 \times 10^6 + 0.15$	$5.0 \times 10^6 + 0.01$	$5.4 \times 10^6 + 0.01$

Values are means of three replicates $\pm$ standard error, *---Significant at P = 0.05
A=Cow dung 5 kg, B=Cow dung 10 kg, C=Cow dung 15 kg, D=Chicken droppings 5 kg, E=Chicken droppings 10 kg, F=Chicken droppings 15 kg, G=Cow dung +Chicken droppings 5 kg, H=Cow dung + Chicken droppings 10 kg, I=Cow dung + Chicken droppings 15 kg, J= Polluted unamended soil, K= Unpolluted soil.

3.2 Mould Counts in Soil Samples after Two Months of Bioremediation

In Table 2, the mould counts were significant in CKD (15 kg) and CD + CKD (10 kg) respectively. Mould counts in chicken droppings have previously been reported to be higher than in cow dung which has been given as a reason for the better performance of chicken droppings as bioremediating agents [12]. However, it can be seen from the table that values were significant ($P=0.05$) after two months of bioremediation.

Table 2. Mould Counts in Soil Samples after Two Months of Bioremediation

Treatment	Mould counts (cfu/g)		
	Time (Months)		
	0	1	2
A	$2.0 \times 10^2 \pm 0.003$	$2.0 \times 10^3 \pm 0.001$	$3.5 \times 10^3 \pm 0.001$
B	$2.2 \times 10^2 \pm 0.001$	$2.2 \times 10^3 \pm 0.001$	$4.0 \times 10^3 \pm 0.001$
C	$1.8 \times 10^3 \pm 0.002$	$5.0 \times 10^3 \pm 0.002$	$5.9 \times 10^3 \pm 0.02$
D	$2.2 \times 10^3 \pm 0.003$	$1.5 \times 10^3 \pm 0.001$	$4.0 \times 10^3 \pm 0.001$
E	$2.5 \times 10^3 \pm 0.01$	$3.3 \times 10^3 \pm 0.002$	$4.6 \times 10^{3} \pm 0.03$
F	$3.0 \times 10^3 \pm 0.001$	$4.0 \times 10^3 \pm 0.001$	$6.0 \times 10^3 \pm 0.01*$
G	$3.5 \times 10^3 \pm 0.003$	$5.5 \times 10^3 \pm 0.03$	$5.0 \times 10^3 \pm 0.004$
H	$3.8 \times 10^3 \pm 0.001$	$5.0 \times 10^3 \pm 0.001$	$6.5 \times 10^3 \pm 0.01*$
I	$4.3 \times 10^3 \pm 0.002$	$5.6 \times 10^3 \pm 0.002$	$5.0 \times 10^3 \pm 0.003$
J	$2.0 \times 10^2 \pm 0.001$	$1.1 \times 10^3 \pm 0.001$	$3.0 \times 10^3 \pm 0.03$
K	$4.0 \times 10^3 \pm 0.01$	$5.5 \times 10^2 \pm 0.001$	$6.0 \times 10^3 \pm 0.02$

Values are means of three replicates $\pm$ standard error, *---Significant at P = 0.05
A=Cow dung 5 kg, B=Cow dung 10 kg, C=Cow dung 15 kg, D=Chicken droppings 5 kg, E=Chicken droppings 10 kg, F=Chicken droppings 15 kg, G=Cow dung + Chicken droppings 5 kg, H=Cow dung + Chicken droppings 10 kg, I=Cow dung + Chicken droppings 15 kg, J= Polluted unamended soil, K= Unpolluted soil.

3.3 Yeast Counts in Soil Samples after Two Months of Bioremediation

For the yeast counts, they were significant in CD (15 kg), CKD (15 kg) and CD + CKD (15 kg) (Table 3). However, no growth of yeasts was detected in CD (5 kg and 10 kg) and POME polluted soil at zero time of bioremediation. This indicates that the POME had a negative impact on the presence of yeasts at the initial time of POME application. There was however, a gradual increase in counts which showed a recovery of the yeasts from the impact of the POME over a period of two months.

Table 3. Yeast Counts in Soil Samples after Two Months Bioremediation

Treatment	Yeast counts (cfu/g)		
	Time (Months)		
	0	1	2
A	No detectable growth	$3.3 \times 10^2 \pm 0.015$	$4.2 \times 10^3 \pm 0.02$
B	No detectable growth	$5.0 \times 10^2 \pm 0.018$	$6.2 \times 10^3 \pm 0.013$
C	$3.0 \times 10^2 \pm 0.001$	$4.7 \times 10^2 \pm 0.01$	$7.9 \times 10^3 \pm 0.03^*$
D	$1.2 \times 10^2 \pm 0.02$	$2.8 \times 10^2 \pm 0.04$	$3.9 \times 10^3 \pm 0.01$
E	$3.4 \times 10^2 \pm 0.01$	$4.3 \times 10^2 \pm 0.002$	$5.3 \times 10^3 \pm 0.03$
F	$4.1 \times 10^2 \pm 0.01$	$7.0 \times 10^2 \pm 0.01$	$6.5 \times 10^3 \pm 0.02^*$
G	$3.5 \times 10^2 \pm 0.02$	$4.2 \times 10^2 \pm 0.02$	$5.4 \times 10^3 \pm 0.01$
H	$4.0 \times 10^2 \pm 0.01$	$5.2 \times 10^2 \pm 0.03$	$6.0 \times 10^3 \pm 0.02$
I	$5.3 \times 10^2 \pm 0.001$	$6.3 \times 10^2 \pm 0.01$	$7.5 \times 10^3 \pm 0.02^*$
J	No detectable growth	$1.3 \times 10^2 \pm 0.01$	$3.0 \times 10^3 \pm 0.01$
K	$1.5 \times 10^3 \pm 0.02$	$1.8 \times 10^3 \pm 0.02$	$1.9 \times 10^3 \pm 0.01$

Values are means of three replicates $\pm$ standard error; *---Significant at P = 0.05
A=Cow dung 5 kg, B=Cow dung 10 kg, C=Cow dung 15 kg, D=Chicken droppings 5 kg, E=Chicken droppings 10 kg, F=Chicken droppings 15 kg, G=Cow dung + Chicken droppings 5 kg, H=Cow dung + Chicken droppings 10 kg, I=Cow dung + Chicken droppings 15 kg, J= Polluted unamended soil, K= Unpolluted soil.

3.4 Occurrence of Bacteria in Amended Palm Oil Mill Effluent (POME) Polluted Soil

In Table 4, *Bacillus subtilis* and *Pseudomonas aeruginosa* had the highest frequency of occurrence of 100% and 91% respectively after two months of bioremediation while *Staphylococcus aureus* occurred least frequently (39.4%) occurring in only two treatments (cow dung, 10 kg and unpolluted soil) at zero time of POME application. *E.coli*, *Proteus vulgaris* and *Micrococcus roseus* had frequencies of 94%, 91% and 76% respectively with *M. roseus* not occurring in POME polluted soil at all throughout the period of bioremediation (0-2 months). *Proteus vulgaris* was absent initially for most treatments but was detected at the end of the bioremediation process for almost all the soil treatments (Table 4). This indicates that it was greatly affected by the acidic nature of the raw POME on application [2]. The presence of *P.vulgaris* up to the second month of bioremediation corroborates the report of [13] in which the hydrocarbon biodegrading potential of *P. vulgaris* in oil polluted sites was reported.

Table 4. Occurrence of Bacteria in Amended Palm Oil Mill Effluent Polluted Soil

Treatment/ Time (months)	Bacterial Isolates																	
	Pseudomonas aeruginosa			Bacillus sp.			Staphylococcus aureus			Escherichia coli			Proteus vulgaris			Micrococcus roseus		
	0	1	2	0	1	2	0	1	2	0	1	2	0	1	2	0	1	2
A	+	+	+	+	+	+	-	+	+	+	+	+	+	+	+	+	+	-
B	+	+	+	+	+	+	+	+	+	+	+	+	-	+	+	-	+	+
C	+	+	-	+	+	+	-	-	-	+	+	+	+	+	+	+	+	+
D	+	+	+	+	+	+	-	-	-	+	+	+	+	+	+	+	+	+
E	+	+	+	+	+	+	-	-	+	+	+	+	+	+	+	+	+	+
F	+	+	+	+	+	+	-	-	+	+	+	+	+	+	+	+	+	+
G	-	+	+	+	+	+	-	+	-	+	+	+	+	+	+	-	+	+
H	+	+	+	+	+	+	-	-	+	+	+	+	+	+	+	-	-	+
1	+	+	+	+	+	+	-	+	-	+	+	+	+	+	+	-	-	+
J	-	+	+	+	+	+	-	-	-	-	-	+	-	+	-	-	-	-
K	+	+	+	+	+	+	+	+	+	+	+	+	+	+	+	+	+	+

A=Cow dung 5 kg, B=Cow dung 10 kg, C=Cow dung 15 kg, D=Chicken droppings 5 kg, E=Chicken droppings 10 kg, F=Chicken droppings 15 kg, G=Cow dung + Chicken droppings 5 kg, H=Cow dung + Chicken droppings 10 kg, I=Cow dung + Chicken droppings 15 kg, J= Polluted unamended soil, K= Unpolluted soil, + = Presence of bacteria, - = Absence of bacteria.

3.5 Occurrence of Moulds in Amended Palm Oil Mill Effluent (POME) Polluted Soil

Table 5 shows the dominance of *Aspergillus* sp. (94%) and *Penicillium verrucosum* (100%) over other fungi isolates. *Rhizopus oryzae* occurred least frequently (27.2%) with it occurring in only two treatments out of the eleven treatments at zero time and after one month of bioremediation. Other moulds such as *Mucor mucedo*, *Fusarium* spp., *Trichophyton* spp. and *Trichoderma harzianum* had frequencies of 69.7%, 87.9%, 69.7% and 63.6% respectively. However, *Rhizopus oryzae* and *Trichoderma harzianum* were not detected in POME polluted soil throughout the period of bioremediation. The moulds isolated from the amended polluted soil in the field were genera of *Rhizopus, Aspergillus, Mucor, Fusarium, Trichophyton, Paecilomyces,* and *Penicillium* (Table 5). This means that these fungi are widespread in the soil [14]. The breakdown of petroleum hydrocarbons by moulds particularly of the genera *Aspergillus, Trichoderma, Penicillium, Mucor* and *Fusarium* has been reported by several authors [12,15,16]. *Aspergillus* sp. in particular are reported to be good producers of cellulases; the enzymes responsible for the breakdown of cellulose in POME [17,18]. Fungi are notably aerobic and can also grow under environmentally stressed conditions such as low pH and poor nutrient status [19,20].

3.6 Occurrence of Yeasts in Amended Palm Oil Mill Effluent (POME) Polluted Soil

In Table 6, the occurrence of yeasts was more frequent in *Candida albicans* (97%) and *Rhodotorula rubra* (81.8%) and least in *Saccharomyces cerevisiae* (45.5%). No growth was observed for *Saccharomyces cerevisiae* at all in two treatments (cow dung, 10 kg and cow dung/chicken droppings, 10kg). Also, *Torulopsis candida* with a frequency of 72.7% was not detected in cow dung/chicken droppings (10 kg) while *Rhodotorula rubra* was not observed at all in unpolluted soil throughout the period of observation. This indicates that it was introduced by POME and the organic wastes. *Rhodotorula* species have however been reported to be oil degraders [12] and in particular, good degraders of anthracene in soil [21]

The ability of the yeasts isolated (*Saccharomyces cerevisiae*, *Torulopsis candida*, *Rhodotorula rubra* and *Candida albicans*) to degrade POME which was demonstrated by their moderate frequency of occurrence (Table 6) in the field has also been reported by [12] though as petroleum utilizers.

Table 5. Occurrence of Moulds in Amended Palm Oil Mill Effluent Polluted Soil

Treatment /Time months	Aspergillus sp.			Mucor mucedo			Penicillium verrucosum			Fusarium sp.			Trichophyton sp.			Rhizopus oryzae			Trichoderma harzianum		
	0	1	2	0	1	2	0	1	2	0	1	2	0	1	2	0	1	2	0	1	2
A	-	-	+	-	+	+	+	+	+	-	-	+	+	+	+	+	-	+	-	+	+
B	+	+	+	-	+	+	+	+	+	+	+	+	+	-	-	-	-	-	-	+	+
C	+	+	+	-	-	+	+	+	+	+	-	+	+	+	+	-	-	+	+	+	+
D	+	+	+	+	-	-	+	+	+	+	-	+	-	+	-	-	-	-	-	+	+
E	+	+	+	+	+	+	+	+	+	+	+	+	-	-	+	-	+	-	-	+	+
F	+	+	+	+	+	+	+	+	+	+	+	+	+	-	+	-	-	+	+	+	+
G	+	+	+	+	-	-	+	+	+	+	+	-	+	-	-	-	-	-	-	+	+
H	+	+	+	+	-	+	+	+	+	+	+	+	+	+	-	-	-	+	-	+	+
I	+	+	+	+	+	+	+	+	+	+	+	+	+	+	+	-	-	+	+	+	+
J	+	+	+	+	+	+	+	+	+	+	+	+	-	+	+	-	-	-	-	-	-
K	+	+	+	+	+	+	+	+	+	+	+	+	+	+	+	+	+	+	-	-	-

A=Cow dung 5 kg, B=Cow dung 10 kg, C=Cow dung 15 kg, D=Chicken droppings 5 kg, E=Chicken droppings 10 kg, F=Chicken droppings 15 kg, =Cow dung + Chicken droppings 5 kg, H=Cow dung + Chicken droppings 10 kg, I=Cow dung + Chicken droppings 15 kg, J= Polluted unamended soil, K=Unpolluted soil, + = Presence of moulds, - = Absence of moulds.

Table 6. Occurrence of Yeasts in Amended Palm Oil Mill Effluent Polluted Soil

Treatment /Time months	Candida albicans			Saccharomyces cerevisiae			Torulopsis candida			Rhodotorula rubra		
	0	1	2	0	1	2	0	1	2	0	1	2
A	+	+	+	-	-	-	-	+	+	-	+	+
B	+	+	+	+	-	-	+	+	+	+	+	+
C	+	+	+	+	-	+	+	+	+	+	+	+
D	-	+	+	+	-	-	+	-	+	+	+	+
E	+	+	+	+	+	+	+	+	+	+	-	+
F	+	+	+	-	+	+	+	+	-	+	+	+
G	+	+	+	+	-	-	+	-	+	+	-	+
H	+	+	+	-	-	-	-	-	-	+	+	+
I	+	+	+	-	+	+	+	-	+	+	+	+
J	+	+	+	-	+	+	+	-	+	+	+	+
K	+	+	+	+	+	+	+	+	+	-	-	-

A=Cow dung 5 kg, B=Cow dung 10 kg, C=Cow dung 15 kg, D=Chicken droppings 5 kg, E=Chicken droppings 10 kg, F=Chicken droppings 15 kg, G=Cow dung + Chicken droppings 5 kg, H=Cow dung + Chicken droppings 10 kg, I=Cow dung + Chicken droppings 15 kg, J= Polluted unamended soil, K= Unpolluted soil, + = Presence of yeasts, - = Absence of yeasts.

4. CONCLUSION

Organic wastes are known to have the ability to improve soil physical properties, buffer the soil and improve aggregate stability and the population of soil microorganisms. This indicates that a combination of the different microorganisms in the organic wastes and the conditions in the field provided conductive environment for growth and production of competent enzymes which helped in the breakdown of the organic compounds contained in the POME.

It is therefore concluded that chicken droppings (at 10 kg and 15 kg/4m^2 plot) and a combination of chicken droppings and cow dung (at 10 kg and 15 kg/4m^2 plot) have the ability to significantly increase microbial populations in palm oil mill effluent (POME) polluted soil thereby stimulating the bioremediation of the polluted soil.

ACKNOWLEDGEMENTS

The authors sincerely acknowledge the support of Mr. Ikechukwu Ogbonnaya and Mr. Stephen Abu of the Soil Science departments of F.U.T. Minna and Kogi State University, Anyigba respectively.

COMPETING INTERESTS

Authors have declared that no competing interests exist.

REFERENCES

1. Poku K. *Small-Scale Palm Oil Processing in Africa*. Technology and Engineering, Food and Agriculture Organization of the United Nations, Rome, Italy; 2002.
2. Ahmad A, Ismail S, Bhatia S. Water recycling from palm oil mill effluent (POME) using membrane technology. Desalination. 2003;157:87-95.
3. Bek-Nielsen C, Singh G, Toh TS. Bioremediation of palm oil mill effluent. In: Proceedings of the Porim International Palm Oil Congress 1999; Istana Hotel, Kuala Lumpur, Malaysia; 1999.
4. Okwute OL, Isu NR. Impact analysis of palm oil mill effluent on the aerobic bacterial density and ammonium oxidizers in a dumpsite in Anyigba, Kogi State. African Journal of Biotechnology. 2007a;6(2):116-119.
5. Adebusoye SA, Ilori MO, Nwaugo VO, Chinyere GC, Inyang CU. Effects of palm oil mill effluent (POME) on soil bacterial flora and enzyme activities in Egbama. Plant Products Research Journal. 2008;12:10-13.
6. Okwute OL, Isu NR. The environmental impact of palm oil mill effluent (POME) on some physicochemical parameters and total aerobic bioload of soil at a dump site in Anyigba, Kogi State, Nigeria. African Journal of Agricultural Research. 2007;2(12):656-662.
7. Oboh BO, Ilori MO, Akinyemi JO, Adebusoye SA. Hydrocarbon Degrading Potentials of Bacteria Isolated from a Nigerian Bitumen (Tarsand) Deposit. Nature and Science 2006;4(3):51-57.
8. Buchanan RE, Gibbons NE. Bergey's Manual of Determinative Bacteriology. 8th ed., Williams and Wilkins Co., Baltimore; 1974.
9. Domsch KH, Gams W. Fungi in Agricultural Soils. 1st Edition, Longman Group Ltd., London, UK; 1970.
10. Barnett JA, Pankhurst, RJ. A New Key to Yeasts. North Holland Publishing Company. Amsterdam, Netherlands; 1974.
11. Statistical Package for Social Sciences, SPSS. Computer package for Windows, Version 19.0; 2010. Available: http:www.spss.com
12. Obire O, Anyanwu EC, Okigbo, RN. Saprophytic and crude oil-degrading fungi from cow dung and poultry droppings as bioremediating agents. Journal of Agricultural Technology. 2008;4(2):81-89.
13. Ceyhan N. Biodegradation of pyrene by a newly isolated *P. vulgaris*. Scientific Research and Essays. 2012;7(1):66-77.

14. Awad AHA. Vegetation: A source of air fungal bio-contaminant. 2005;21:53-61.

15. Ibiene AA, Orji FA, Ezidi CO, Ngwobia CL. Bioremediation of hydrocarbon contaminated soil in the Niger Delta using spent mushroom compost and other organic wastes. Nigeria Journal of Agriculture, Food and Environment. 2011;7(3):1-7.

16. Eze VC, Owunna ND, Avoaja DA. Microbiological and Physicochemical Characteristics of Soil receiving palm oil mill effluent in Umuahia, Abia State, Nigeria. Journal of Natural Science Research. 2013;3(7):163-169.

17. Wong KM, Nor AA, Suraini A, Vikineswary S, Mohd AH. Enzymatic hydrolysis of palm oil mill effluent solid using mixed cellulases from locally isolated fungi. Research Journal of Microbiology. 2008;3(6):474-481.

18. Mohanram S, Amat D, Choudhary J, Arora A, Nain L. Novel perspectives for evolving enzyme cocktails for lignocellulose hydrolysis in biorefineries. Sustainable Chemical Process. 2013;1:1-15. doi:10.1186/2043-7129-1-15

19. Davis JB, Westlake DWS. Crude oil utilization by fungi. Canadian Journal of Microbiology. 1979;25:146-156.

20. Frey-Klett P, Burlinson P, Deveau A, Barret M, Tarkka M, Sarniguet A. Bacterial-Fungal Interactions: Hyphens between Agricultural, Clinical, Environmental and Food Microbiologists. Microbiology and Molecular Biology Reviews. 2011;75(4):583-609. doi:10.1128/MMBR.00020-11.

21. Krivobok S, Miriouchkhine E, Seigle-Murandi F, Benoit-Guyod JL. Biodegradation of Anthracene by Soil Fungi. Chemosphere. 1998;37(3):523-530.

Use of Palm Oil Mill Effluent as Medium for Cultivation of *Chlorella sorokiniana*

Charles Ogugua Nwuche[1,2*], Doris Chidimma Ekpo[1],
Chijioke Nwoye Eze[1], Hideki Aoyagi[2] and James Chukwuma Ogbonna[1,2]

[1]*Department of Microbiology, Faculty of Biological Sciences, University of Nigeria, Nsukka 40001, Nigeria.*
[2]*Department of Bioscience and Bioengineering, Graduate School of Life and Environmental Sciences, University of Tsukuba, Ibaraki, Japan.*

Authors' contributions

This work was carried out in collaboration between all authors. Author CON was the project leader and was responsible for the project plan, experimental design, data analyses and writing the manuscript. Authors DCE and CNE collected the samples and contributed to the data analyses. Authors HA and JCO are senior specialists who provided resources for the execution of the study and edited the manuscript. All authors read and approved the final manuscript.

ABSTRACT

Aims: Palm oil mill effluent (POME) erodes the principal biophysical characteristic of both soil and water when discharged untreated but could be exploited as medium for microalgae cultivation due to its vast mineral contents.
Place and Duration of study: POME samples were collected from a local palm oil processing mill at Nsukka, Enugu State, Nigeria. A part of the study was done at the Graduate School of Life and Environmental Sciences, University of Tsukuba, Japan while the rest at the University of Nigeria, Nsukka between March and September, 2012.
Methodology: *Chlorella sorokiniana* C212 was grown in several Batches (A-D) of POME supplemented with urea (60 mg/L) before subjecting to different sterilization protocols. Cultivation was conducted in shaker flasks at 150 rpm, 1 vvm, 3000 lux and pH 7.0 ± 0.2.
Results: The filter sterilized Batch (B) promoted the highest (1070 ± 30 mg/L) dry cell weight (DCW), lipid (156 ± 12 mg/g-cell) and chlorophyll (1.59 ± 0.11 mg/g-cell) contents while chemical oxygen demand (COD) decreased by $45\pm08\%$. The autoclaved medium

Corresponding author: Email: charles.nwuche@unn.edu.ng

(Batch A) gave the least DCW (310±20 mg/L), lipid production (40±05 mg/g-cell) and chlorophyll content (0.58±0.02 mg/g-cell) while COD reduced by 20±04%. The highest COD decrease (70±05%) was achieved in the unsterilized Batch (D). Batch B was most positively affected by dilution because at 75% concentration, DCW increased to 1360±30 mg/L, lipid contents to 174±10 mg/g-cell, chlorophyll to 1.87±0.14 mg/g-cell the while COD declined by 63±03%.
Conclusions: POME has potential for use in microalgae cultivation with significant saving in treatment costs.

Keywords: Palm oil mill effluent; microalgae; cultivation; chlorella sorokiniana; sterilization; chlorophyll; chemical oxygen demand; dilution.

1. INTRODUCTION

Palm oil mill effluent (POME) has received considerable attention in recent years due to its capacity for considerable environmental damage when discharged untreated into inland waterways and cultivated lawns. POME is produced during palm oil production and is a viscous, acidic brown liquid that is predominantly organic with a highly unpleasant odour. In Nigeria, POME is discharged into the environment in its raw form by small scale operators which constitute the major proportion of palm oil refiners. During palm oil extraction from the fresh fruits, about 50% of the water results in POME. It is estimated that for 1 tonne of crude palm oil that is produced, 5 to7.5 tonnes of water ends up as POME [1]. According to the produce department of the Ministry of Agriculture, Kogi State, Nigeria, approximately 3,600 metric tonnes of palm oil is produced annually [2].

Currently interests are growing in the use of POME as feedstock for the production of valuable products and biochemicals because the presence of high concentrations of carbohydrate, proteins, nitrogenous compounds, lipids and minerals in POME [3] make it an excellent raw material for bioconversion by biotechnological means. Presently, one area of application is the cultivation of microalgae which can be processed into biofuels to support the world's continually growing energy demand and curtail the ecological dangers associated with the use of fossil fuel as well as provide a ready environmentally friendly alternative to fossil based energy in the reality of depleting global reserves.

Microalgae are minute photosynthetic organisms that are cultivated on a large scale because they supply many active biological products which are beneficial to man [4]. Apart from biofuel, microalgae can be processed into dietary proteins [5], vitamins, carotenoids, antioxidants [6], fatty acids, enzymes, polymers, toxins and sterols [7]. They are typically found in freshwater and marine system where they have sufficient access to light, water, carbon dioxide and inorganic salts [8]. The major components of media used for photoautotrophic cultivation of microalage include inorganic nitrogen sources and both macro and micro elements. POME contains vast range of mineral elements and can be exploited for the cultivation of microalgae.

This presents an important way of further treating POME before discharge into the environment. Thus as the biomass increase, the organic load of the effluent decreases due to metabolism and uptake of the POME components by the algae. Although a few authors have reported on the use of POME as medium for the cultivation of microalgae, none hitherto has considered the impact of sterilization patterns on biomass and metabolite production nor has the influence of the culture on the kinetics of COD of the effluent been adequately highlighted. In this study the potential of using POME as medium for cultivation of Chlorella sorokiniana was investigated.

2. MATERIALS AND METHODS

2.1 Palm Oil Mill Effluent (POME)

Fresh raw POME used as medium for cultivation of Chlorella sorokiniana C 212 was obtained from a local palm oil processing mill at Nsukka, Enugu State of Nigeria and stored at 4°C until use. POME for use in experiment was first passed through a double-layered muslin cloth (previously sterilized by autoclaving) to remove oil, plant fibres, broken shells and kernels before filtration through Whatman No.1 filter paper. The filtrate was supplemented with 60 mg/L of urea and then sub-divided into several batches. Batch A was sterilised by autoclaving at 121°C for 15 min. Batch B was treated by passing through a Millex GV 0.22 μm filters (Millipore SA, France). Batch C contained POME mixed with chloramphenicol at a final concentration of 50μg/ml while Batch D was used without any additional treatment. Batch E was the conventional BG 11 medium and served as control. It was inoculated after sterilizing for 15 min at 121°C. The treatment which gave the highest cell growth was used to determine the effect of POME concentrations on the kinetic parameters. Identical culture conditions were implemented in a separate batch of POME diluted with distilled water to 25, 50 and 75% concentrations. Except for the BG 11 medium, the optical density of all the POME batches and dilutions were determined against a distilled water blank before inoculation in order to evaluate the degree of light penetration in the media. The media were assessed at 430 and 662 nm which is the absorption maxima for chlorophyll a as well as at 453 and 642 for chlorophyll b.

2.2 Microalgae and Culture Conditions

Chlorella sorokiniana C212 was initially propagated in a 500 ml Erlenmeyer flask containing BG 11 medium at 29±2°C. The cultivation condition was maintained at an aeration rate of 1 vvm and 150 rpm for 7 days. Continuous illumination at 3000 lux was provided by two fluorescent lamps arranged in parallel on both sides at an equidistance of 10 cm from the flask. The cells were harvested by centrifugation at 3000 rpm for 10 min and the supernatant discarded. The cells (pellets) were washed twice with distilled water before they were inoculated into the different POME batches at 10% inoculum concentration equivalent to 4.0×10^8 cells/ml. Approximately 90 ml of each of the different POME treatments were aseptically introduced into 500 ml Erlenmeyer flasks and cultivated as previously described for 21 days. All glass-wares were previously washed, air-dried and sterilized in a hot air-oven at 160°C for 2 h before use. The chemicals and reagents used were reagent grade and purchased from Wako Pure Chemical Ind., Osaka, Japan.

2.3 Analytical Methods

2.3.1 Cell growth

The growth and concentration of the microalgae during batch cultivation was measured spectrophotometrically at 680 nm and converted to biomass by a calibration curve corresponding to the POME treatment or medium used. Specific growth rate (μ) was calculated using equation from the exponential growth:

$$\text{Specific Growth Rate } (\mu) = \ln(x_2) - \ln(x_1)/t_2 - t_1 \tag{1}$$

Where: x_2 and x_1 are the concentrations of the biomass at the end and beginning of each batch run, while t_2 and t_1 is the duration of the run.

2.3.2 Dry cell weight

At the end of cultivation, biomass was harvested by centrifugation at 4000 rpm for 15 min and the supernatant discarded. The pellets were washed with distilled water and freeze-dried at -52°C under vacuum. The dry biomass obtained after freeze-drying was stored in airtight containers at 20°C while cell weight was determined gravimetrically in mg/L.

2.3.3 Chlorophyll content

Chlorophyll *a* content was determined spectrophotometrically at 665 nm wavelength in methanol (90%) extracts of the dried biomass according to a modification of Lee and Shen [9] equation:

$$\text{Chlorophyll } a \text{ (g/g-cell)} = 13.43 \times OD_{665} \tag{2}$$

2.3.4 Lipid content

The lipid content of the microalgal biomass was estimated by a modified method of Bligh and Dyer [10]. The total lipids were extracted by a mixture of chloroform-methanol (2:1 v/v) and the biomass at the ratio of 1:1. The mixture was allowed to stand for 1 h before they were transferred into a separatory funnel, shaken for 5 min and allowed to settle. The lipid fraction was then separated and the solvent evaporated in a rotary evaporator. The weight of the crude lipid was then obtained gravimetrically using a digital balance.

2.3.5 POME characteristics

The pH of the different POME treatment and concentrations were determined according to Standard Methods [11]. Chemical Oxygen Demand (COD), Ammonia-Nitrogen (NH_3-N), Total Nitrogen (N) and Nitrates (NO_3) were measured by the Hach's Spectrophotometic method (DR/4000, Hach Co. Ltd. Tokyo). Carbohydrate content was analyzed by the phenol–sulphuric acid method [12] while soluble protein was measured by the Bradford method [13].

2.4 Statistical Analysis

The data is presented at the means of triplicate experiments ± standard deviations. Where appropriate, results were statistically analyzed by the completely randomized one-factor analysis of variance (ANOVA) using the statistical software IBM-SPSS (Statistical Product and Service Solutions) 16.0.2 (2008 Version).

3. RESULTS

3.1 Effect of POME Treatment on Biomass Production

Biomass production varied in the different POME batches according to the sterilization protocol applied Table 1. In the batch sterilized by autoclaving (Batch A), the DCW of the algae was 310±20 mg/L while in Batch C (sterilized by chloramphenicol (50µg/ml) treatment), it was 640±15 mg/L. Algal cultivation was conducted in Batch D without initial medium sterilization but the total DCW produced at the end of cultivation was 458±27 mg/L. The Table 1 reveals that the POME treatment most beneficial to biomass production (1070±30 mg/L) was the Batch (B) sterilized by membrane filtration. However, the overall result indicate that biomass obtained from the Batch E (2945±90 mg/L) was about three times the DCW from Batch B and several folds higher than the values obtained from the other Batches. When the algae was cultivated in different concentrations Table 2 of a separate batch of the filter sterilized POME, results show that the DCW of biomass (1360±30 mg/L) harvested from the 75% POME was higher than those obtained from the other concentrations. Statistical analysis showed that the differences were statistically significant (P=0.05). At lower concentrations (50 and 25%), the DCW were 874±40 and 560±25 mg/L respectively. The biomass productivity pattern reflects the trend described for the DCW obtained from the different media. The productivity rate was highest (210.36±6.43 mg/L/d) in cultures cultivated in the conventional medium but among the POME treatments, the rate obtained from the Batch B (76.40±2.15 mg/L/d) was the highest while Batch A (22.14±1.42 mg/L/d), the least. In Batches C and D, biomass grew at the rates of 45.71±1.07 and 32.71±1.93 mg/L/d respectively.

3.2 Effect of POME Treatment on Chlorophyll Production

The chlorophyll content of biomass harvested from the different media shows Table 1 that cultures grown in the control medium (BG 11) had the highest chlorophyll content (2.98±0.12 mg/g-cell). Among the POME batches, the concentration of the pigment was statistically higher (P=0.05) in biomass harvested from Batch B (1.59±0.11 mg/g-cell) than those from batch C (1.07±0.15 mg/g-cell). In the other Batches, chlorophyll extracted from Batch a (0.58±0.02 mg/g-cell) was higher than values (0.41±0.07 mg/g-cell) obtained from those grown in Batch D. The chlorophyll content of biomass from the different POME dilutions Table 2 also showed that pigments from the 75% POME (1.87±0.14 mg/g-cell) was higher than those obtained from the other dilutions. In the 100% POME, for instance, the chlorophyll content was 1.59±0.11 mg/g-cell but at lower concentrations, 0.85±0.18 and 0.48±0.07 mg/g-cell of chlorophyll was obtained from the 50 and 25% POME respectively. In terms of productivity, cultures grown in the BG 11 medium (Batch E) had the highest rate (0.213±0.01 mg/L/d) followed by those from Batches B (0.114±0.01 mg/L/d) and C (0.076±0.01 mg/L/d). Chlorophyll productivity in cultures grown in Batch a (0.041±0.00 mg/L/d) and Batch D (0.029±0.01 mg/L/d) were not different (P>0.05) statistically. At lower POME concentrations, cultures grown in the 75% POME had values (0.134±0.01 mg/L/d) significantly higher than those obtained from other Batches Table 1 and dilutions Table 2. The productivity of cultures from the 50% POME (0.061±0.01 mg/L/d) was also higher (P=0.05) than the rates obtained from the 25% POME (0.034±0.01 mg/L/d).

Table 1. Kinetic parameters of C. sorokiniana C212 grown in BG 11 medium and different POME treatments

POME	Cell Dry Weight (mg/L)	Chlorophyll Content (mg/g-cell)	Lipid Content (mg/g-cell)	Biomass Productivity (mg/L/d)	Specific Growth Rate (μ)	Chlorophyll Productivity (mg/L/d)	Lipid Productivity (mg/L/d)	% COD Reduction (mg/L)
Batch A	310±20*	0.58±0.02*	40±05*	22.14±1.42*	0.4098*	0.041±0.00*	2.86±0.36*	20±04
Batch B	1070±0*	1.59±0.11*	156±12*	76.43±2.15*	0.4982*	0.114±0.01*	11.1±0.86*	45±08
Batch C	640±15*	1.07±0.15*	94±25*	45.71±1.07*	0.4615*	0.076±0.01*	6.7±1.79*	58±03
Batch D	458±27*	0.41±0.07*	52±07*	32.71±1.93*	0.4376*	0.029±0.01*	3.71±0.5*	70±05
Batch E∞ (BG 11)	2945±90	2.98±0.12	386±18	210.36±6.43	0.5706	0.213±0.01	27.6±1.29	**

*Asterisk indicates significant difference from the control treatment (N = 5, P = 0.05, mean SE from ANOVA); **COD of the BG 11 medium was not determined; ∞ Batch E is Control Treatment

Table 2. Kinetic parameters of C. sorokiniana C212 grown in different concentrations of the filter-sterilized (Batch B) POME

POME Conc. (%)	Cell Dry Weight (mg/L)	Chlorophyll Content (mg/g-cell)	Lipid Content (mg/g-cell)	Biomass Productivity (mg/L/d)	Specific Growth Rate (μ)	Chlorophyll Productivity (mg/L/d)	Lipid Productivity (mg/L/d)	% COD Reduction (mg/L)
75	1360±30*	1.87±0.14*	174±10*	97.14±2.14*	0.5153*	0.134±0.01*	12.4±0.72*	63 ±03*
50	874±40*	0.85±0.18*	120±10*	55.30±7.12*	0.4838*	0.061±0.01*	8.57±0.72*	45±04
25	560±25*	0.48±0.07*	90±08*	40.00±1.80*	0.4520*	0.034±0.01*	6.42±0.57*	28±04*

* Asterisk indicate significant difference from Batch B values (N = 4, P = 0.05, mean SE from ANOVA)

3.3 Effect of POME Treatment on Lipid Production

The lipid content of biomass harvested from the control medium was 386±18 mg/g-cell. This value was higher than results obtained from the other POME treatments Table 1. The lipid profile of biomass from Batch B (156±12 mg/g-cell) and Batch C (94±25 mg/g-cell) was different (P=0.05) statistically. As earlier indicated, 75% POME was found to be the best medium for lipid accumulation (174±10 mg/g-cell) among cultures grown in the different POME dilutions. Lipid content of the 50% POME was 120±10 mg/g-cell but it reduced to 90±08 mg/g-cell in the 25% POME. Lipid productivity followed the same trend as the earlier results. Among the different batches, Batch B stimulated the highest lipid productivity (11.1±0.86 mg/L/d) while Batch A, the least (2.86±0.36 mg/L/d). At lower POME concentrations, 75% POME promoted the highest lipid productivity (12.4±0.72 mg/L/d) while the 25% had the least (6.42±0.57 mg/L/d). Biomass cultivated in the conventional medium had the highest rate (27.6±1.29 mg/L/d) of lipid accumulation.

3.4 Effect of POME Treatment on COD Changes

COD changes in the different POME batches show Table 1 that the least rate (20±04%) of COD reduction was achieved in the POME treatment sterilized by autoclaving (Batch A) while the highest rate (70±05%) occurred in the unsterilized POME medium. The rate of COD reduction in the batches sterilized by filtration (B) and antibiotic treatment (C) were 45±08 mg/L and 58±03 mg/L respectively. Changes in COD profiles of the different POME concentrations were in line with the trend observed in other results. The rate of COD reduction in the 75% POME (63±03%) was higher than the rates found in the other concentrations. At 50 and 25% POME, COD declined by 45±04% and 28±04% respectively.

4. DISCUSSION

The POME batch treated by membrane filtration (Batch B) promoted the highest biomass (1070±30 mg/L) DCW among the different batches tested. This is thought to be due to the concentration of light available to the microalgae during cultivation. At all the wavelengths and among all the batches tested Table 4, the optical density (OD) of Batch B was found to be the lowest indicating that light penetration was maximum compared to Batch A (autoclaved) which had the least light availability as shown by the high OD values. The OD indicates the degree of clarity of the media. The latter correlate to amount of transmitted light and directly affects biomass production by photosynthesis. At the two maximum absorption regions (430 and 662 nm) for chlorophyll a for instance, the OD of Batch A were 1.476 and 0.897 while values for the filter sterilized POME (B) were 0.168 and 0.072. This evidently implies that light was more available to the microalgae in Batch B than in the other treatments. Due principally to this advantage, biomass increased remarkably to a final concentration (1070±30 mg/L) which was significantly (P=0.05) higher than values obtained from the other batches.

Microalgae are sunlight-driven organisms capable of converting carbon dioxide to biomass and other high value products [4,14]. One other factor that is noteworthy is the chemistry of the medium following filter-sterilization. Millipore preserves the physical, chemical and biological quality of materials and is known to be effective against both bacterial and fungal contamination [15]. This implies that the various components of the POME were retained after the filtration and available to the cultures in their original and natural forms. The effect of urea was also found to be more prominent in this medium than in the other batches.

Nitrogen has been identified as a primary nutrient for the growth of several microalgae [16]. Urea was selected as a nitrogen supplement in this study based on preliminary evaluation Fig. 1 which ascertained that urea was better (1.47 g/L) than the other inorganic salts in stimulating higher biomass production by the algae. This observation was in agreement with Su et al. [17]. In the present study, the addition of nitrogen salt was justified due to the relatively low content of the element in POME. Table 3 shows that the concentration of nitrate in the raw POME was only 40±05 mg/L while other nitrogen (total and ammoniacal) was 90±04 and 109±07 mg/L respectively. At higher POME dilutions, the concentration of the element further diminished. Urea enhanced the growth of the algae by promoting maximum optical density and shorter lag phases in line with the findings of Hadiyanto and Nur [18]. The modifying effect of urea on pH of the effluent is also worthy of note. The initial pH (4.02±0.07) increased to 7.0±0.2 on addition of urea indicating potential savings in cost as no further chemical was applied for pH adjustment. The final pH was favourable to the growth and metabolism of the algae and meets the discharge criterion for industrial effluents.

Table 3. Characteristics of the different POME concentrations used in the study

Parameters (mg/L)	100%**	75%	50%	25%
pH*	3.97±0.03	3.95±0.05	3.98±0.04	4.04±0.05
Protein	44.2±8.0	35.1±6.5	28.2±4.0	15±5.0
COD***	8200±00	6000±00	3940±00	1790±00
Total Carbohydrate	584±20	490±25	354±15	264±10
Ammonia Nitrogen (NH_3-N)	109±07	75±04	56±03	25±02
Total Nitrogen (N)	90±04	62±02	46±05	21±04
Nitrates (NO_3)	40±05	27±03	20±04	09±04

*Not measured in mg/L, ** Data in column indicate characteristic ± standard deviation of the different POME batches. *** Values represent the COD of the filter-sterilized (Batch B) POME. COD for batches A, C and D was 9160±120*

Table 4. Optical density of the different POME batches and concentrations before inoculation at the absorption maxima for chlorophylls *a* and *b*

POME	Chlorophyll *a*		Chlorophyll *b*	
	430 (nm)	662 (nm)	453 (nm)	642 (nm)
Batch A	1.476	0.897	1.378	0.930
Batch B	0.168*	0.072*	0.129*	0.075*
75%	0.143*	0.062*	0.113*	0.071*
50%	0.099*	0.051*	0.080*	0.053*
25%	0.074*	0.046*	0.063*	0.047*
Batch C	1.083*	0.550*	0.987*	0.580*
Batch D	1.083*	0.550*	0.987*	0.580*

Asterisk indicates significant difference from the Batch A values (N = 7, P = 0.05, mean SE from ANOVA)

Biomass from Batch C was 640±15 mg/L. Although the antibiotic may be effective against susceptible bacterial populations in the medium but the presence of other organisms, including the resistant flora, fungi and their metabolites could be inhibitory to the growth of the algae despite the presence of adequate light and culture conditions. Perhaps, the most likely occurrence of interference by other microbial groups could be seen in cultures grown in the unsterilized POME (458±27 mg/L). The least DCW was found in the POME batch sterilized by autoclaving Table 1. The heat treatment changed the colour of the medium to blue-black. The exact reason for the change in colour is not clear but it is thought that some chemical reactions involving tannic acid and other components occurred during exposure to the high temperature of steam sterilization. The presence of tannic acid and its involvement in the darkening of POME has been confirmed [19]. The darkening led to shading which limited light penetration in the medium and negatively affected chlorophyll formation.

Chlorophyll traps the radiant energy that powers photosynthesis [20]. Its concentration is also known to correlate with photosynthetic activity and directly influences the production of biomass and the accumulation of target products [21]. This is in line with results obtained in the present study. Findings from Table 1 show that the high chlorophyll content (1.59±0.11 mg/g-cell) of the cells cultivated in Batch B was directly related to the increase in biomass (1070±30 mg/L) and lipid content (156±12 mg/g-cell) of the microalgae. The reverse was equally true in the case of the other batches with lower chlorophyll concentrations. Cultures grown in the conventional (BG11) medium had a chlorophyll content of 2.98±0.12 mg/g-cell which positively affected the overall outcomes of the DCW (2945±90 mg/L) and lipid content (386±09 mg/g-cell). Mineral salt media are usually not as complicated as industrial waste waters. POME contains numerous organic and inorganic compounds [22] as well as toxic components which occur at concentrations that are highly variable and often subject to a wide range of influences [23]. The BG11 is an adequate and nutritionally balanced medium that supplies the right concentrations of essential nutrients and minerals needed for the best possible growth of freshwater algae [24]. This explains the positive impact of the medium on the various kinetic components of the culture as indicated by the high specific growth rate (0.5706).

Evaluation of DCW from the different POME concentrations Table 2 showed that the 75% POME stimulated higher biomass production (1360±30 mg/L) than the 100% POME (1070±30 mg/L) as well as the lower POME concentrations. This contrasts previous study by Hadiyanto et al. [25] which reported higher biomass productivity in the 50% POME. The wide differences in POME composition [23] as well as the concentration of urea added (1 g/L) in the referred article could account for this variation. The lower growth of biomass in the raw POME could be attributed to some toxic components. The presence of phenols and some organic acids is generally believed to be responsible for the phytotoxic and antibacterial activity of POME [26-27]. The growth of Chlorella was poor in the 25% POME as shown by the low specific growth rate (0.4520) despite the lowered concentration of the toxic elements and higher light intensity. Such low growth rate could be caused by over dilution of the key nutrients [25]. Evaluation of the rate of COD reduction by the algae under the different culture conditions indicate that Chlorella could be exploited as potential candidate for the treatment of POME. The algae grown in Batch A recorded a very low (20±04%) reduction rate due to identified reasons but the COD of the Batch B and C cultures declined by 45±08 and 58±03% respectively. The former represent the metabolic capacity of a homogenous algal monoculture, while the latter suggests the involvement of other microorganisms besides the algae by the higher reduction values. This is confirmed from cultures grown in unsterilized POME (Batch D) in which the highest rate of COD reduction (70±05%) was achieved despite having the least (0.41±0.07 mg/L) chlorophyll content.

In cultures grown in the different concentrations of Batch B POME Table 2, COD reduction was enhanced (63±03%) especially in the 75% POME. This further lends credence to the possibility of interference by some components of the raw POME as indicated by the increase in the rate of COD reduction at a lower (75%) concentration. However, despite the limitations imposed by the inhibitory substances, the organisms present in Batch D (unsterilized) appeared unaffected and even achieved a higher rate of COD reduction

(70±05%) than the other cultures. Monocultures usually metabolize a limited range of substrates but a mixed microbial consortium as may be present in the unsterilized POME can be more efficient in remediation due to their broad enzymatic capacities and higher tolerance to fluctuations of temperature, pH and salinity [28].

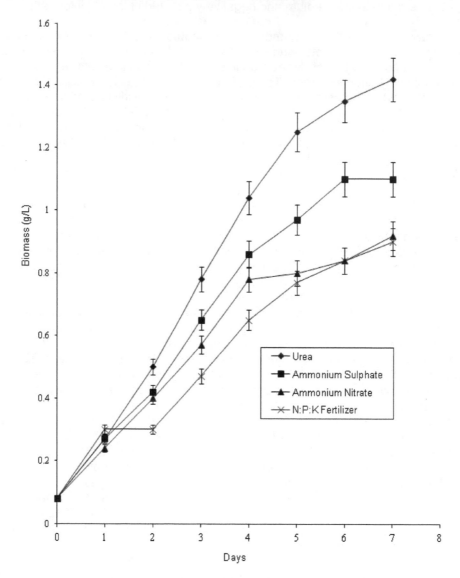

Fig 1. Effect of different nitrogen compounds on the biomass of *C. sorokiniana*

5. CONCLUSION

POME has potential for use as medium for microalgal cultivation with significant saving in treatment costs. However, the low nitrogen content of POME implies that additional supplementation with appropriate compound is needed to improve the nutritional quality of the effluent. Also, due to variation in the composition of POME, the concentration that promotes the best biomass or product yield has to be pre-determined in order to reduce the negative impact of the inhibitory components on the metabolism of cultures.

COMPETING INTERESTS

Authors have declared that no competing interests exist.

REFERENCES

1. Ahmad AI, Ismad S, Bhatia S. Water recycling from palm oil mill effluent (POME) using membrane technology. Desal. 2003;157:87-95.
2. Okwute OL, Isu NR. Impact analysis of palm oil mill effluent on aerobic bacterial density and ammonium oxidizers in a dump-site at Anyigba, Kogi State. Afr J Biotechnol. 2007;6:116–119.
3. Habib MAB, Yusoff FM, Phang SM, Ang K, Mohammed S. Nutritional values of chironomid larvae grown in palm oil mill effluent and algal culture. Aquaculture. 1997;158:95–105.
4. Singh S, Bhushan NK, Banerjee UC. Bioactive compounds from cyanobacteria and microalgae: an overview. Crit Rev Biotechnol. 2005;25:73–95.
5. Ogbonna JC, Masui H, Tanaka H. Sequential heterotrophic / autotrophic cultivation: an efficient method of producing Chlorella biomass for health food and animal feed. J Appl Phycol. 1997;9:359–366.
6. Munro MHG, Blunt JW, Dumdei EJ, Hickford SJH, Lill RE, Battershill CN, Duckworth AR. The discovery and development of marine compounds with pharmaceutical potentials. J Biotechnol. 1999;70:15–25.
7. Pulz O, Gross W. Valuable products from biotechnology of microalgae. Appl Microbiol Biotechnol. 2004;65:635–648.
8. Chisti Y. Biodiesel from microalgae. Biotechnol Adv. 2007;25:294–306.
9. Lee YK, Sheen H. Basic culturing techniques. In: Richmond A, editor. Handbook of microalgal culture: Biotechnology and applied phycology.UK. Blackwell Science Ltd; 2004.
10. Bligh EG, Dyer WJ. A rapid method of total lipid extraction and purification. Can J Biochem Physiol. 1959;37:911–917.
11. APHA, AWWA and WPCF. Standard methods for the examination of water and wastewater. 21st Edition, American Public Health Association, Washington DC., USA; 2005.
12. DuBois M, Gilles KA, Hamilton JK, Rebers PA, Smith F. Colorimetric method for determination of sugars and related substances. Anal Chem. 1956;28:350–356.
13. Bradford MM. A rapid and sensitive method for the quantitation of microgram quantities of protein using the principles of protein – dye binding. Anal Biochem. 1976;72:248–254.
14. Spolaore P, Joannis-Cassan C, Duran E, Isambert A. Commercial applications of microalgae. J Biosci Bioeng. 2006;101:87–96.
15. Moliterni E, Jimenez-Tusset RG, Rayo MV, Rodriguez L, Fernandez FJ, Villasenor J. Kinetics of biodegradation of diesel fuel by enriched microbial consortia from polluted soils. Int J Environ Sci Technol. 2012;9:749–758.
16. Wen YW, Chen F. Optimization of nitrogen sources for heterotrophic production of eicosapentanoic acid by the diatom Nitzschia laevis. Enzyme Microb Technol. 2001;29:341–347.
17. Su CH, Giridhar R, Chen CW, Wu WT. A novel approach for medium formulation for growth of a microalga using motile intensity. Biores Technol. 2007;98:3012–3016.

18. Hadiyanto H, Nur MMA. Potential of palm oil mill effluent (POME) as medium growth of Chlorella sp for bioenergy production. Int J Environ Bioenergy. 2012;3:67-74.

19. Phalakornkule C, Mangmeemak J, Intrachod BN. Pre-treatment of palm oil mill effluent by electrocoagulation and coagulation. Science Asia. 2010;36:142–149.

20. McIntyre HL, Kana TM, Anning T, Geider RJ. Photoacclimation of photosynthesis irradiance response curves and photosynthetic pigments in microalgae and cyanobacteria. J Phycol. 2002;38:17–38.

21. Su CH, Fu CC, Chang YC, Nair GR, Ye JL, Chu IM, Wu WT. Simultaneous estimation of chlorophyll a and lipid contents in microalgae by three colour analysis. Biotechnol Bioeng. 2008;99:1034–1039.

22. Habib MAB, Yusoff FM, Phang SM, Kamarudin MS, Mohammed S. Chemical characteristics and essential nutrients of agro-industrial effluents in Malaysia. Asian Fish Sc. 1998;11:279–286.

23. Mercade ME, Manresa MA, Robert M, Espuny MJ, Andres C, Guienea J. Olive oil mill effluent (OOME): New substrate for biosurfactant production. Biores Technol. 1993;43:1–6.

24. Stanier RY, Kunisawa R, Mandel M, Cohen-Bazire G. Purification and properties of unicellular blue-green algae (order Chroococcales). Bacteriol Rev. 1971;35:171–205.

25. Hadiyanto H, Nur MMA, Hartanto GD. Cultivation of Chlorella sp as biofuel sources in palm oil mill effluent (POME). Int J Renew Energy Dev. 2012;1:45–49.

26. Capasso R, Cristtinzio G, Evidente A, Scognainiglio F. Isolation spectroscopy and selective phytotoxic effects of polyphenols from vegetable wastewaters. Phytochem. 1992;31:4125–4128.

27. Pascual I, Antolin AC, Garcia C, Polo A, Sanchez-Diaz M. Effect of water deficit on microbial characteristics in soil amended with sewage sludge or inorganic fertilizer under laboratory conditions. Biores Technol. 2007;98:29–37.

28. Boopathy R. Factors limiting bioremediation technologies. Biores Technol. 2000;74:63–67.

Identification of Bacterial Population of Activated Sludge Process and Their Potentials in Pharmaceutical Effluent Treatment

Farrokhi Meherdad[1], Ghaemi Naser [2], Najafi Fazel[3], Naimi- Joubani Mohammad[1], Farmanbar Rabiollah[4] and Roohbakhsh Joorshari Esmaeil[1*]

[1]*Environmental Health Department, School of health, Guilan University of Medical Sciences, Rasht, Iran.*
[2]*Biochemistry & Biotechnology Department, Tehran University of Medical Sciences, Tehran, Iran.*
[3]*School of base of science, Guilan University, Rasht, Iran.*
[4]*Health Education Department, School of health, Guilan University of Medical Sciences, Rasht, Iran.*

Authors' contributions

This work was carried out in collaboration between all authors. Authors FM and RJE designed the study. Authors FR and GN, performed the statistical analysis. Authors FM and RJE wrote the protocol. Authors NF, NJM and RJE wrote the first draft of the manuscript. Authors RJE and NJM managed the analyses of the study. Authors NF and RJE managed the literature searches. All authors read and approved the final manuscript.

ABSTRACT

Aims/objectives: The cognition about microbial population of activated sludge and their treatment potential will be very useful for industrial wastewater treatment plant operation.
Methodology: In this study microbial population of activated sludge process that was used for pharmaceutical wastewater has been investigated. Sampling was done from return sludge line and after serial dilution 1500 plates were studied. Methods for separating the bacteria from wastewater was pour plate method. All bacterial samples were purified using nutrient Agar and Macconkey Agar culture. Bacteria were separated from return sludge line and classified into 3 groups after biochemical tests and

---

Corresponding author: Email: Esmaeil5115@yahoo.com

morphological analysis, These include positive bacteria of Bacillus genus, Pseudomonas aeruginosa and Flavobacterium.
Results: The biodegradability study on pharmaceutical effluent using identified cultures in laboratory scale showed that Bacillus spp. are the most efficient bacteria for organic matter degradation.
Conclusion: Results of this study showed that providing a microbial bank of these spp. can be useful for resistant operation of activated sludge.

Keywords: Pharmaceutical wastewater; microbial population; activated sludge; COD.

1. INTRODUCTION

Nowadays one of the most important environmental contaminants is pharmaceuticals. It is expected that their worldwide production will increase, and substantial amounts of pharmaceuticals can reach the environment, either through direct discharge into the water resources or due to the inefficient removal in wastewater treatment plants [1]. There is little similarity between characteristics of wastewater from different pharmaceutical factories. Commonly these effluents contain little biological oxygen demand (BOD) and it can be negligible but contain higher chemical Oxygen Demand (COD) [2].

The Biological processes are commonly used for industrial wastewater because these methods are economical and environmental sound. Most industrial wastewaters contain multi-component mixture of various types of organics [3,4]. These organics can be broadly classified as, readily degradable or bio-refractory. Certain pharmaceuticals wastewater contains a large amount of bio-refractory and toxic compound [4]. Therefore biological treatment of pharmaceuticals wastewater will be very difficult and Conventional biological treatment plants are usually inappropriate for the treatment of pharmaceutical effluent [5].

When treating industrial wastewaters, particularly for removal of toxic or refractory organics, it is necessary acclimate the biomass to wastewater and conservation of acclimated biomass specially, bacteria are considered as an important action [5].

Identification of bacterial population as the most important organisms for wastewater treatment can be very useful for stable operation of wastewater treatment plant. Aerobic and anaerobic bacteria are used for pharmaceutical wastewater treatment [6]. Many studies on pharmaceutical wastewater treatment plant shows that the microbial population is vary from one plant to another and wastewater characteristics and environmental factors affecting the microbial population [5,6]. It has been found in some cases the acclimated biomass to possess a genetic memory [7]. This is of great importance for those industries with intermittent production of specific products such as pharmaceutical industry.

The main aims of this study were: to identify the bacteria that are involved in pharmaceutical wastewater treatment in an activated sludge process, to determine their efficiencies for COD removal and determine the most efficient bacteria and provide a bacterial bank for seeding to the activated sludge process in emergencies condition such as hydraulic, organic and toxic shocks.

2. MATERIALS AND METHODS

2.1 Chemicals

All chemicals were purchased from Merck and Aldrich companies in analytical grade.

2.2 Experimental Methods

2.2.1 Microbial culture

The first step in screening was providing pure culture or mixed population with the highest removal efficiency. Sampling was done for 1 month from return sludge line and sent to laboratory in a safe and standard manner. The volume of each sample was 4 liter. The samples were diluted by serial dilution method to 7 dilutions (10^1, 10^2, ...10^7). All diluted sample were cultured and incubated at 37°C by spread plate and pour plate method with about 1500 plate evaluated. To make pure culture, each plate was sub-cultured in 3 tubes containing nutrient agar.

2.2.2 Determination of treatment potential of bacteria

Potential of activated sludge for pharmaceutical wastewater treatment was first determined using 48 hr. In this method 4 reactors with volume of 10 liter were selected and aerated by a blower (Fig.1). The mixture of sterilized pharmaceutical wastewater and activated sludge in the reactors 1, 2, 3, and 4 were as shown in Fig 1. For evaluation of treatment potential of different bacterial cultures (3 identified cultures) for pharmaceutical wastewater, 5 days aeration method was used. This method was conducted same as 48 hr method except that 5 day aeration and pure culture instead of activated sludge were used.

2.3 Analytical Methods

All tests and analytical methods conducted according to the standard methods for the examination of water and wastewater [6].

3. RESULTS AND DISCUSSION

Table 1 shows the specification of bacteria in dilution of 10^5. It can be seen so many colonies are growth on both of culture nature with different specification.

Table 2 shows the classification of identified bacteria as morphological and biochemical characteristics. Table 3 show the results of biochemical tests.

The results of biochemical tests revealed that, it can be concluded that the MS 3-1 is *Pseudomonas aeruginosa*, the $P2M10^{-4}2$ is *Flavobacterium* spp, and the $P1N10^{-3}3$ is *Bacillus* spp. with their distribution in the activated sludge as *Pseudomonas aeruginosa* (50%), *Flavobacterium* spp. (13.5%), and bacillus spp. (1.19%) [8]. These microbes can be maintained and exploited for efficient maintenance and operation of wastewater treatment plant [9,10]. These results conformed to earlier studies [11,12]. Moreover, results of this study indicated the antibacterial effects of engineered nanomaterials implications for wastewater treatment plantsThe *Pseudomonas aeruginosa* spp Specific baced D.G.G.E (denaturing gradient gel electrophoresis) [13].

Table 1. Specification of bacteria in dilution of 10^5

No	specification culture nature		Number of colonies	Apparent specification of colonies	Number of bacteria	Number of gr^+	staining
1	N.A	First plate	150	Dispersed milky colonies	5×10^7	33	gr^+ gr^-
		Second plate	125	Milky dispersed colonies some smooth and some hard	12.5×10^7	6	gr^+ gr^-
2	Mac .A	First plate	17	Silver and red dispersed colonies and aerobic	11.7×10^7	-	gr^-
		First plate	19	Dispersed pink colonies and aerobic	11.9×10^7	0	gr^-

gr^- : Gram negative; gr^+ :Gram positive;   N.A: Nutrient agar;   Mac. A: macconkey agar

Table 2. Classification of bacteria from morphological and biochemical points of view

No	Oxidative-fermentative test	Cotalas test	Oxides test	Form and characteristics of bacteria	Specification of colonies on nutrient agar and Mac cankey agar	specification vcxzVCZcgf
1	+	+	+	gr^-	Blue colonies on	MS-3-1*
	+	+	+	Coccus and diplo coccus gr^-	nutrient agar	M4-2P
2	+	+	+	gr^-	Fine to medium	$P2m10^{-4}$ 1
	+	+	+		brown	$P1m10^{-4}$ 1
	+	+	+		Medium brown	$P1m10^{-3}$ 2
	+	+	+		Clear chocolate brown	$P2m10^{-3}$ 2
					Geometrical cream colonies	
3	+	+	+	gr^-	Big brown	$P2m10^{-4}$ 3
	+	+	+		mucous	$P2m10^{-4}$ 2*
	+	+	+		chocolate	$P1m10^{-4}$ 2
	+	+	+		Brown-chocolate	$P1N10^{-3}$ 1
					Brown to violate	
					Small brown mucous	
4	+	+	-	gr^+	Yellow colonies 1-2 ml on nutrient agar	$P1N10^{-3}$ 3

+ : Positive;   - : negative; gr^- : Gram negative; gr^+ :Gram positive;
Gr^- Coccus: Coccus Gram negative   diplo coccus gr^- : diplo coccus Gram negative

Table 3. Biochemical tests for identification of MS 3-1, P2M10^{-4}2 and P1N10^{-3}3 bacteria

Type of Bacteria	Morphological	Negative	Positive
MS 3-1	Gr $^-$ double coccus	glucose, lactose, sucrose, SH$_2$,indole , Co$_2^+$H$_2$, VP , Of with oil	MR, Citrate, oxides, Of, catalase, Malonate, Motility
P2M10^{-4}2	Bacillus gr $^-$	Lactose ,sucrose, SH$_2$ indole , Co$_2^+$H$_2$,VP, Motility	MR , glocuse,Citrate, oxidase ,Of, catalase, Malonate, Of with oil
P1N10^{-3}3	bacillus gr $^+$ spore forming	SH$_2$, indole, Co$_2^+$H$_2$,VP Motility, Malonate , oxides ,Of	MR, catalase, citrate Of with oil

Methods were developed and important in many WWTPS (Wastewater treatment plants). (Please check grammar of this sentence) Results of this study indicated that pharmaceuticals reduced diversity of microbial community of activated sludge [15]. The results of biochemical tests have confirmed that the MS 3-1 is *Pseudomonas aeruginosa*, the P2M10^{-4}2 is *Flavobacterium* spp, and the P1N10^{-3}3 is Bacillus spp. As a result of Lapra et.al, research, percents of identified bacteria in activated sludge was *Pseudomonas aeruginosa* (50%), *Flavobacterium* spp. (13.5%), and *Bacillus* spp. (1.19%) [8]. According to the study of Rani et al., the most microbial diversity in activated sludge was related to bacillus and pseudomonas. These microorganisms help us to reach efficient maintenance and operation in wastewater treatment plant [9]. Also this results with study [11,12] Moreover Results of this study indicated the antibacterial effects of engineered nanomaterials implications for wastewater treatment Plants [13,14]. Results of this study indicated that pharmaceuticals reduced diversity of microbial community of activated sludge [15]. (Are these sentences replicated? Please check it).

Fig. 1 shows the decreasing of COD after 3, 6, 12, 24 and 48 hr were reported as treatment efficiencies.

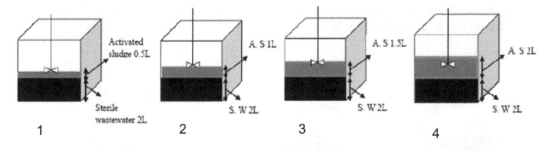

Fig. 1. Wastewater treatment steps in lab scale using activated sludge

Fig. 2 Shows the potential of activated sludge for treating pharmaceutical waste water in different concentration. The more concentration of activated sludge is seeded, the more efficiency of COD reduction is achievable.

It can be seen that after 24 hr the COD reduction is constant and increasing the aeration time have no effect on COD decreasing, therefore it can be resulted that optimum detention time of aeration tank is 24 hr [8,16].

These results are consistent with results of LaPara et al. [8]. That showed the COD removal was about 62% at 30°C and the extent of soluble COD removal declined as temperature increased by an average of 60 mg/L per °C [5]. (Please check grammar of this sentence)

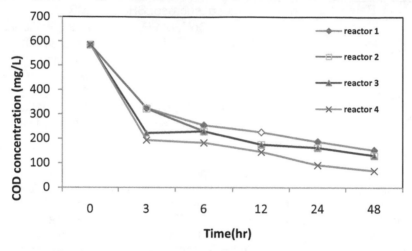

Fig. 2. The potential of activated sludge for removal of COD

Fig. 3 shows the experimental set up for evaluation of treatment potential of identified bacteria for pharmaceutical wastewater.

- As shown in Fig. 4, *Bacillus* spp. is the most efficient bacteria for pharmaceutical wastewater treatment. However no significant differences among the efficiencies of three bacteria spp (PV=0.99). It is not consistent with results of Madukasi [17] that reports *Rhodobacter spheroidies* Z08 has proven to be effective in ameliorating hazardous pollutants found in pharmaceutical wastewater with over 80% COD reduction [5]. It can be seen when pure culture are used for wastewater treatment the optimum aeration time increased (2 days) in comparison to mixed population as activated sludge. The reason is related to role of other organisms in wastewater treatment such as fungi, protozoa, etc.

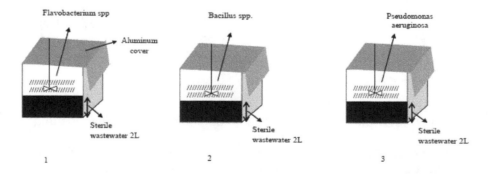

Fig. 3. Wastewater treatment steps in lab scale using isolated bacteria

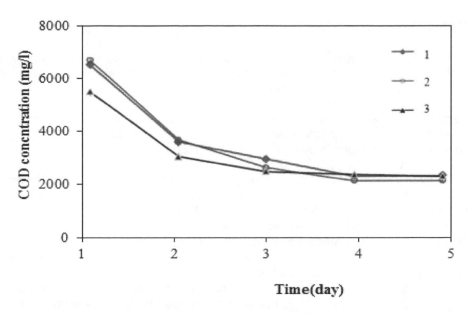

Fig. 4. The COD removal efficiencies of each of identified bacteria in pharmaceutical wastewater treatment.

4. CONCLUSION

Results from this study indicate that the pharmaceutical wastewater can be treated by activated sludge within 24 hr aeration time.

Three species of bacteria are identified as following

1- *Flavobacterium* spp.
2- *Bacillus* spp.
3- *Pseudomonas aeruginosa*

Bacillus spp. were the most efficient bacteria for pharmaceutical wastewater treatment, and providing a microbial bank of this spp can be useful for resistant operation of activated sludge and prevention of effects of toxic, organic and hydraulic shocks.

ACKNOWLEDGEMENT

I would like to express my special thanks to the Guilan university of medical science - Faculty of Health officials for Financial and moral support of this project.

COMPETING INTERESTS

Authors have declared that no competing interests exist.

REFERENCES

1. De los Reyes FL, Ritter W, Raskin L. Group-Specific small-subunit rRNA hybridization probes to characterize filamentous foaming in activated sludge systems. Applied Environ. Microbiol.1997;63:1107-17.
2. Bitton G. Waste water Microbiology. 3rd ed., Wiley-Liss, Inc: New York, USA; 2005.
3. Manz W, Wagner M, Amann R, Schleifer KH. In situ characterization of the microbial consortia active in two wastewater treatment plants. Water Res.1994;28:1715-23.
4. Farrokhi M, Mesdaghinia AR. Removal of 3-monochlorophenol in anaerobic baffled reactor. J. Applied Sci. 2007;7:1652-55.
5. Sekiguchi Y, Kamagata Y, Syutsubo K, Ohashi A, Harada H, Nakamura K. Phylogenetic diversity of mesophilic and thermophilic granular sludges determined by 16S rRNA gene analysis. Microbiology.1998;144:2655-65.
6. APHA, AWWA, WEF. Standard Methods for the Examination of Water and Wastewater. 22nd ed. American Water Works Association, USA; 2005.
7. Thompson G, Forster C. Bulking in activated sludge plants treating paper mill wastewaters. Water Res. 2003;37:2636-44.
8. LaPara TM, Nakatsu CH, Pantea LM, Alleman JH. Aerobic biological treatment of a pharmaceutical wastewater: Effect of temperature on COD removal and bacterial community development. Water Res. 2001;35:4417-25.
9. Rani A, Porwal S, Sharma R, Kapley A, Purohit HJ, Kalia VC. Assessment of microbial diversity in effluent treatment plants by culture dependent and culture independent approaches. Bioresource Technol. 2008;99:7098-07.
10. Urbain V, Block JC, Manem J. Bioflocculation in activated sludge: An analytic approach. Water Res.1993;27:829-38.
11. Tambosi JL, Yamanaka LY, Jose HJ, Moreira RFPM, Schroder HF. Recent research data on the removal of pharmaceuticals from Sewage Treatment Plants (STP). Química Nova. 2010;33:411-20.
12. Amna A, Fozia N. Frecuency distribution of bacteria isolated from different industrial effluents. Daffodil Int. Univ. J. Sci. Tecnol. 2012;7:28-33.
13. Musee N, Thwalaa M, Nota N. The antibacterial effects of engineered nanomaterials: Implications for wastewater treatment plants. J. Environ. Monit. 2011;13:1164-83.
14. Mao Y, Zhang X, Yan X, Liu B, Zhao L. Development of group-specific PCR-DGGE fingerprinting for monitoring structural changes of *Thauera* spp. in an industrial wastewater treatment plant responding to operational perturbations. J. Microbiol. Methods. 2008;75:231-36
15. Kraigher B, Kosjek T, Heath E, Kompare B, Mandic-Mulec I. Influence of pharmaceutical residues on the structure of activated sludge bacterial communities in wastewater treatment bioreactors. Water Res. 2008;42:4578-88.
16. Farrokhi M, Ebrahimpur M, Amirmozafari N, Isazadeh KH, Naimi Joubani M, Omidi S. Comparison of the efficiency of municipal and industrial activated sludge for hospital wastewater treatment. J. Guilan Univ. Med. Sci. 2012;22:9-14.
17. Madukasi EI, Dai X, He C; Zhou J. Potentials of phototrophic bacteria in treating pharmaceutical wastewater. Int. J. Environ. Sci. Technol. 2010;7:165-74.

9

Different Methods for DNA Extraction from Yeast-*Candida famata* Isolated from Toddy

T. Santra[1], S. K. Ghosh[2*] and A. Chakravarty[3]

[1]Department of Microbiology, Institute of Genetic Engineering, Badu, Kol-128, India.
[2]Molecular Mycopathology Lab, P. G. Department of Botany, Ramakrishna Mission Vivekananda Centenary College, Rahara, Kol-118, India.
[3]Department of Molecular biology, Institute of Genetic Engineering, Badu, Kol-128, India.

Authors' contributions

This work was carried out in collaboration of all authors. Author TS did the extraction procedure of DNA. Author SKG designed the biochemical characterization, managed the analysis of whole study and wrote the manuscript, and author AC managed the literature search. All authors read and approved the final manuscript.

ABSTRACT

Aims: Isolation and biochemical characterization of yeasts from toddy and standardization of best method for DNA extraction from yeast.
Study Design: Biochemical characterization of yeast and genomic DNA extraction by manual and kits methods.
Place and Duration of Study: Department of Microbiology, Institute of Genetic Engineering, Badu, kol-128, India and Molecular Mycopathology Lab, P. G. Department of Botany, Ramakrishna Mission Vivekananda Centenary College, Rahara, kol-118, India, from November 2012 –April,2013.
Methodology: Toddy was collected in sterilized polythene bags from palm tree (*Borassus flabellifer* L; Family: *Arecacea*) in the morning, from Badu, 24-parganas (N) India. Isolation of yeasts was done by the method of Beech and Davenport [15] using MA (Malt extract) medium. Biochemical Identification was performed by using basal medium and procedure [1,2,15]. Genomic DNA extraction was done by manual and kits methods (Uniflex[TM] DNA isolation Kit). Quality of extracted DNA was checked by the absorbance ratio (A_{260} / A_{280}) ranged from 1.8 to 2.0.
Results: By performing morphological, microscopical and biochemical characterization

Corresponding author: Email: swapan.krghosh@yahoo.com

the isolated yeast from toddy was identified as *Candida famata* consulting with the key of yeast published [1,2].

The Uniflex[TM] DNA isolation Kit method is much more convenient way to get pure and high quality DNA than the manual methods.

Conclusion: Isolated yeast from toddy was identified as *Candida famata*. The genomic DNA of *Candida famata* was extracted purely by Uniflex[TM] DNA isolation Kit. This method was better and more convenient than manual method.

Keywords: Toddy; Yeast; biochemical characterization; DNA extraction; Kit method.

1. INTRODUCTION

Yeasts are unicellular fungi. They may be Ascomyceteous, Basidiomyceteous or Deuteromyceteous fungi. Some yeasts are filamentous in specific environmental conditions. They are called dimorphic yeasts. Yeasts are cosmopolitan. They are generally present in natural sources such as leaf surface, fruit surface, soil, flowers, various juices (toddy juice, date palm juice), molasses, animal surfaces, etc [1]. They are generally saprophytic but some are also pathogenic to humans. Therefore, their diversity in natural sources is very immense. According to Kurtzman and Fell [2], there are about 100 genera and 700 species of yeasts. In most advanced countries like USA, Japan and Germany systematic study of yeast flora from natural sources has taken much emphasis in biocontrol of pathogens. Although, India including West Bengal has vast natural sources for yeast, its study of yeast flora is limited.The post harvest rots of fruits and vegetables are chronic problem in post harvest technology i.e fruit industry. The major post harvest rot of orange (*Citrus reticulate* Blanco) is caused by *Penicillium digitatum or P. italicum.* Potential use of yeasts as biocontrol agents of soil born fungal plant pathogens and as plant growth promoters were recent investigated by EL- Tarabily and Sivasiyhamparam [3] and have been used extensively for biological control of post harvest diseases of fruits & vegetables[4].The yeast *Torulapsis candida (Candida famata)* effectively controls *Penicillium digitatum* infection on *Citrus* fruits [5].Wild variety of yeasts have Can*dida saitoana and C. oleophila* control post harvest diseases of apple and *Citrus* fruits[6].*Candida famata* gave maximum percent of radial inhibition of growth (70.24 PIRG)followed by *Pichia membranifaciens* (68.21 PIRG), *Rhodotorula mucilaginosa* (60.56 PIRG)[7]. *Candida famata* (also known as *Debaryomyces hansenii* and *Torulopsis candida*) is a commensal yeast found in cheese, dairy products and the environment. *C. famata* is now an emmerging human pathogen and it accounts for 0.2%–2% of invasive candidiasis of human[8].The cell wall is the main obstacle for quick and easy lysis of yeasts and therefore it must be disrupted for efficient recovery of genomic DNA (gDNA). Conventional methods for gDNA preparation from yeast cells utilize either enzymatic degradation generally followed by lysis of cells with detergent and extraction of gDNA with phenol-chloroform. When analyzing large number of samples these methods are time consuming and relatively expensive procedure for extracting DNA. Various methodology have been reported for DNA extraction [9,10,11]. Two simple and easy protocols for extracting the high quality DNA from yeast have been employed in this work. One protocol is done by preparing the lysis buffer, organic solvent and dissolving buffer manually and the second was the specific kit (Uniflex[TM] DNA isolation Kit) method. Pure DNA extraction is very necessary for routine genotyping of yeasts either by simple detection of PCR products or RAPD(Random Amplified Polymorphic DNA), or for initial amplification of genomic DNA for sequencing; procedures that are widely used for analysis of scientific, environmental, industrial and clinical samples.

The main objective of this study was to screen the best method of these two procedures to extract pure and high quality DNA of C.famata and getting the convenient, easy, less time consuming method for the yield of high quality and pure DNA having the absorbance ratio (A_{260} / A_{280}) ranged from 1.8 to 2.0.

2. MATERIAL AND METHODS

2.1 Isolation and Purification Procedure of yeast

Toddy was collected in sterilized polythene bags from palm tree (*Borassus flabellifer* L;Family: *Arecaceae*) in the morning, from Badu, 24-parganas (N) India and brought in laboratory. One ml of toddy was serially diluted in sterile distilled water upto 10^{-4} and these were plated on sterile MA medium (MA; Malt extract, 2g;Agar,2g; Distilled water,100ml in sterile petridish and incubated for 48 hours at 28 ± 2°C in BOD incubator [2,12].

Single cell of the isolate was obtained by streaking loop full of cells on MA medium and transferring well isolated colonies to MA. The isolates were maintained on Malt-Yeast – Glucose- Peptone –Agar medium (Dry malt extract, 3g;dry yeast ,3g; peptone,5g; D-glucose, 10g; agar,20g; distilled water, 1L) at 28±2°C with monthly subculturing.

2.2 Identification of Yeast

Identification of isolated yeast up to species level was carried out on the basis of standard morphological, and physiological /biochemical tests [1,2,13,14].

2.2.1 Morphological and Microscopical investigation

The colonies were observed and described on MA and MYGPA medium. The isolated yeast was also grown in MA and MYPG broth for determination of their cultural characteristics (pellicle, sedimentation or ring formation). In certain cases, the isolate was grown on sterile slices of carrot for induction of ascospore formation.

The isolated yeast grown on MYPG broth and its slide was prepared and stained by crystal violate solution and observed under oil emersion lens (100xX10x) of compound light microscope.

2.2.2 Biochemical characterization

For carbon and nitrogen assimilation, the basal medium [1,15] was used and the results were determined after the 3th, 7th, 14th, 21th and 28th day.

The ability of some carbohydrates for anaerobic assimilation (fermentation) was determined by using Durhan glass tubes after 3 weeks. The quantity of the tested carbohydrates was 2%. For Diazonium blue –B (DBB) test, a ten day old culture on MYPGA was held at 55°C for three hours and then flooded with ice –cold DBB reagent. The reagent was prepared by dissolving diazonium blue salt (Sigma) in cold 0.5M-tris –HCL buffer pH 7.0 at 1mg /ml.The reagent was kept ice –cold and used within few minutes of preparation. Other additional tests such as starch formation, urea hydrolysis, cyclohexamide (0.01% or 0.1%), were performed [1].

2.3 DNA Extraction Method

2.3.1 Extraction of DNA from yeast culture by preparing the buffer in laboratory

Overnight grown (at 30°C) yeast culture (1x10^7 cells /ml) was taken in an centrifuge tube. Yeast culture was centrifuged for 5 min at 1500 rpm and it was resuspended in 0.5 ml of sterile distilled water. Cells were transferred to eppendorf and it was spin down for 5 sec. at 14,000 rpm. Supernatant was discarded and pellet in the residual water was vortexed 200 µl of yeast lysis buffer (Triton X -100, 10% SDS, 5 M NaCl, 0.5 MEDTA, 1M Tris and distilled water), 200 µl of organic solvent (phenol: chloroform : isoamyl alcohol in the ratio of 25:24:1), and 0.3g of glass beads were added. It was again vortexed and 200 µl TE was added. It was centrifuged again and aquous phase was transferred to a new tube and 1 ml of ethanol was added. Supernatant was discarded after centrifugation and 400 µl TE (Tris-EDTA) and 4 µl RNase was added to the pellet and kept at 37°C. 10 µl of 4M ammonium acetate and 1ml ethanol was added. Again it was spin and supernatant was discarded. It was washed with 70% ethyl alcohol, then pellet was air dried and resuspended in 50 µl TE. Agarose gel (1%) electrophoresis was done and band was observed under UV transluminator. Purity was checked by taking the absorbance ratio in spectrophotometer. This was done according to the published keys [9,10,11].

2.3.2 Uniflex^{TM} DNA isolation Kit method

Overnight grown (at 30°C) yeast culture (1x10^7 cells /ml) was taken in an centrifuge tube. Yeast culture was centrifuged for 5 mins at 6000 rpm. 1 ml of uniflex^{TM} buffer 1 and 10 µl of RNase A were added to the pellet , mixed well and kept for 30 mins at 37°C for incubation. 1 ml of 1:1 phenol: chloroform was mixed to the lysed cell. It was centrifuge at 10,000 rpm for 15 mins at rt [room temperature]. The aquous layer was transfered to a fresh tube and uniflex^{TM} buffer 2 was added and mixed well. It was again centrifuge at 12,000 rpm for 15 mins and supernatant was discarded. 70% ethanol was added to the pellet and centrifuged at 10,000 rpm for 10 mins and supernatant was discarded. The pellet was air dried and resuspended in 50 µl UNIFLEX^{TM} elution buffer and it was stored at -20°C. Agarose (1%) gel electrophoresis was done and band was observed under UV transluminator. Purity was checked by taking the absorbance ratio in spectrophotometer.

3. RESULTS AND DISCUSSION

The data presented in the Table 1 showed that the isolated yeast colony morphology was white and round; its margin was undulating but elevation was convex. The microscopical study revealed that each cell was elliptical, budding present but no ascospore and no ballistospore. All these morphological and microscopical characteristics were at per with characteristics of *Candida famata* shown in the key of Barnetts et al. [1].

Table 1. Morphological and microscopical characteristics of yeast isolate

Characteristics	Yeast Isolate
Colour	White
Surface	Round
Margin	Undulated
Elevation	Convex
Cell- shape	Ellipsoid.
Ascospore	Absent
Ballistispore	Absent
Pseudomycelium	Absent
True mycelium	Absent

The biochemical characterization of the isolated yeast isolate shown on Table 2, indicated that this isolate gave positive result (growth) to D-glucose, D-galactose, L-sorbose, D xylose, sucrose, maltose, trehalose, cellobiose, lactose, rafinose, inulin, glecerol, D-mannitol of carbon assimilatory test while negative (no growth) to starch, myoinositol, 2 keto-D-glucose, D-glucoronate, succinate, citrate, methanol of carbon assimilatory test. It did fermented in maltose but in other carbon sources fermentation was weak or not happened. This isolate did not grow in nitrate or shown delayed or weak growth in nitrite and L- lysine. Regarding the effect of temperature, it could growth from 25 -35°C while it did not growth at temperature from 37°C and above. This isolate was sensitive to 0.01% of cyclohaxamide.

All these biochemical characteristics compared with the standard keys of yeasts [1,2,13,14] suggested that the yeast isolated from toddy palm belongs to *Candida famata* [Fig. 1].

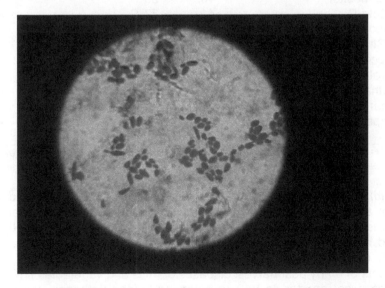

Fig. 1. Microscopic view of yeast isolate (*C. famata*) (10xX100x).

Table 2. Biochemical characteristics of yeast isolate using carbon assimilating tests and other tests

Sr. No.	Carbon assimilation test	Yeast Isolate	Sr. No	Carbon assimilation test	Yeast isolate
1.	D-glucose	+	26	Butane 2,3-diol	N
2.	D-galactose	+		Nitrogen assimilating tests	
3.	L-sorbose	+	27	Nitrate	-
4.	D-xylose	+	28.	Nitrite	D,W
5.	D-arabinose	D ,W	29.	L-Lysine	D,W
6.	L-ramnose	-	30.	Cadaverine	N
7.	Sucrose	+	31.	Glucosamine	N
8.	Maltose	+		Fermentation in carbon source	
9.	α,αTrehalose	+	32.	D-glucose	W
10.	Cellobiose	+	33.	D-galactose	-
11.	Lactose	+	34.	Sucrose	-
12.	Raffinose	+	35.	Maltose	+
13.	Inulin	+	36.	Lactose	-
14.	Starch	-	37.	Inulin	N
15.	Glycerol	+		Growth at different temperatures	
16.	D-glucitol	+	38.	25°C	+
17.	D-mannitol	+	39.	30°C	+
18.	Myo-inositol	-	40.	35°C	+
19.	2-Keto-D-glucose	-	41.	37°C	-
20.	D-Glucuronate	-	42.	40°C	-
21.	Succinate	-		Different Tests	
22.	Citrate	-	43.	Diazonium blue -B	-
23.	Methanol	-	44.	Urea hydrolysis	-
24.	Ethanol	D	45.	Cyclohexamide (.01%	-
25.	Propane 1,2- diol	N	46.	Cyclohexamide (0.1%)	-
			47.	Starch formation	-

Responses: +=Positive growth;   - = Negative (no growth) ;W=weak growth ;w/-weak or negative; W/+= weak or positive N=Not determined; D= Delayed growth ;VW=Very weak growth ;

Simillarly *C.famata* and other species of this yeast *(C. tropicalis, C.krusei* and *C.valida*) were isolated from toddy palm by other workers [7,16]. *Candida* sp are anamorphic yeasts. *Candida* sp was isolated and characterized from palm syrup, molasses, toddy and grapes in India [17] and moreover, *Candida famata* was isolated from the fruit surface of *Syzygium cumini* L [12]. *Candida famata* was reported as antagonistic to *Penicillium digitatum* and biocontrol agents of many post harvest diseases of fruits and vegetables [18,19,20]. Recently *Candida famata* was reported to be one emerging human pathogen and some yeasts are emerging opportunistic animal pathogen [21]. Desnos-Ollivier et al.[22] reported *C. famata* is a rare human pathogen. *C. famata,* while frequently isolated from air, soil, water, plant material and animals, and human and animal faeces, has been found to be a very rare aetiological agent in disease processes in animals and humans. A case of lifelong episodic diarrhoea in a dog that might have been aggravated by colonisation of the intestinal mucosa by *S. cerevisiae* and *C. famata* is reported [23]. In animals it has been isolated from the udder of cows with mastitis, genital secretions of ruminants, the mouth of vitamin A-deficient pigs, and a fungal arthritis of the fetlock in a horse [24]. Therefore, *Candida famata*, are being recognized as potential biocontrol agent of post harvest diseases of fruits and vegetables and on the other hand it is being emerging pathogens that cause several types of infections in humans and animals. Under this situation very quick and perfect the identification of this fungus is urgent. In our experiments for identification of yeast, all morphological, microscopical and biochemical tests took one week. The total time required for identification of *Candida* species using species-specific PCR is less than 5h including 2h for DNA extraction, 1h and 40 min for PCR and 40min agarose gel electrophoresis. It is considered rapid as the identification can be done within a working day, as compared to conventional biochemical tests which require more than 5 days [25]. Detection and identification of fungal DNA by PCR is one of the most powerful and popular tools for the early detection and identification of pathogenic fungi, including *Candida* species [26]. In this sense, extraction and purification of DNA of *Candida famata* and protocol for its quick molecular or PCR identification are very much necessary.

L 1 of the Fig.2 showed the band obtained from manual method. This band was poor. L3 and L7 of the Fig.2 indicated the band obtained from DNA ladder and kit method respectively. The band of L7 (Kit method) was very prominent, bright and its size was more greater than 3000 kb as indicated by DNA ladder. The amount of gDNA extracted by kit method is 62ng /10 µg while it was 35ng/10µg by manual method. It was done by Quantifying soft wire (Biorad). Opara et al. [27] also obtained good result using kit method for DNA extraction.

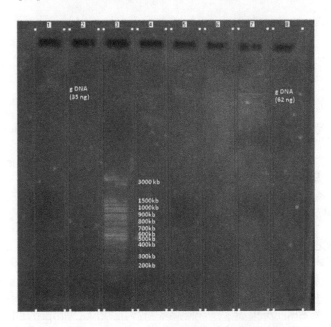

Fig. 2. DNA band observed from manual methodology (L1), from Kit method (L7) and DNA ladder (L3)

Cell lysis is the main step for efficient recovery of DNA. Nucleic acids must be solubilized from cells or other biological material. Conventional method for DNA extraction from yeast cells mainly include cell lysis and either enzymatic degradation or treating with glass beads. This method requires addition of precipitation solution (ethyl alcohol) and TE buffer in two times which slows down the whole protocol and extraction of large number of DNA was not properly handled. So, this method is time consuming and it is inconvenient way to extract the high quality DNA. This was done according to the published keys [27].

On the contrary the kit (Uniflex[TM] DNA isolation Kit) method belongs to basic DNA extraction protocol which include cell lysis, enzymatic treatment, phenol-chloroform and precipitation. The purity of DNA was checked by spectrophotometric method and the absorbance ratios (A_{260}/A_{280}) obtained ranged from 1.8 to 2.0. The absorbance value obtained by kit method was 1.79 where as the value obtained by manual method was 1.5.

So this kit method is much more convenient way to get pure and high quality DNA than the manual method.

4. CONCLUSION

Toddy is one of the most important habitat of yeast, particularly *Candida famata* and this yeast is highly antagonistic to *Penicillium digitatum* early reported; on the other hand it is an emerging pathogen of animals. Isolated yeast from toddy was identified as *Candida famata*. The genomic DNA of *Candida famata* was extracted purely by Uniflex[TM] DNA isolation Kit. This method was better and more convenient than manual method. This kit method for extraction of genomic DNA would be very useful for molecular or PCR identification of this yeast or other yeasts .

COMPETING INTERESTS

Authors have declared that no competing interests exist.

REFERENCES

1.   Barnett JA, Payne RW, Yarrow D. Yeasts characteristics and identification. 3rd ed. Cambridge University Press.2000.

2.   Kurtzman CP, Fell JW. The yeasts – a taxonomic study, North Holland, Amsterdam.1999;1055.

3.   EL-Tarabily KA, Sivasiyhamparam K. Potential of yeasts as biocontrol agents of soil borne fungal plant pathogens and as plant growth promoters. Mycoscience 2006;47: 25-35.

4.   Punja ZK. Comparative efficacy of bacteria, fungi and yeasts as biological control agents for disease of vegetable crops. Can J Plant Pathol.1997;19:315-323.

5.   Arras G, Dessi R, Sanna P, Arru S. Inhibitory activity of yeasts isolated from fruits against *Penicillium digitatum*. Acta. Horticulturae. 1999;48(5):37-45.

6.   El- Ghaouth A, Smilanick JL, Wisniewski M, Wilson CL. Improved control of apple fruit decay with a combination Candida saitona and 2-deoxy-D-glucose.Plant Disease,2000;84(2):248-253.

7.   Ghosh SK, Santra T, Chakravarty A. Study of antagonistic yeasts isolated from some natural sources of West Bengal. Agric. Biol. J. N. America. 2013;4(1):33-40.

8.   Nicholas D. B, Chuang SH, Alam MJ , Shah DN, Ng TM, McCaskey L, Garey KW. Treatment of *Candida famata* bloodstream infections: case series and review of the literature J. Antimicrob. Chemother. 2012; doi: 10.1093/jac/dks388

9.   Lõoke M, Kersti K, Arnold K. Extraction of genomic DNA from yeasts for PCR- based applications Biotechniques. 2011;50(5):325–328. doi:10.2144/000113672.

10.   Ling M, Merante F, Robinson BH. A rapid and reliable DNA preparation method for screening a large number of yeast clones by polymerase chain reaction. Nucleic acids research. 1995;23:4924–4925.

11.   Harju S, Fedosyuk H, Peterson KR. Rapid isolation of yeast genomic DNA: Bust n' Grab. BMC Biotechnology. 2004;4:8.

12.   Ghosh SK. Study of yeast flora from Fruit of *Syzyzium cunini*. Agri. Bio. J N. America. 2011;2(8):1166-1170.

13.   Kurtzman CP, Fell JW, Boekhout T. The yeasts – a taxonomic study. 5th Ed. North Holland, Amsterdam. 2012.

14.   Ross A, Harrisson JS. The Yeast. Academic Press, London. 1987-1993; Vol. 1-5.

15.   Beech FW, Davenport RR. Methods in Microbiology. 1971;4:153-182.

16.   Tuntiwongwanich S, Leenanon B. Morphology and identification of yeasts isolated from toddy palm in Thailand J. Micro .Soci. of Thailand. 2009;23(1):34-37.

17.   Ghosh SK, Samaddar KR. Characterization and biomass production potential of yeast flora of some Natural sources of Kalyani. J. Mycopathol. Res. 1991;29 (2):111-117.

18.   Droby S, Wisniewski M, El-Ghauoth A, Wilson C. Biological control of post harvest diseases of fruits and vegetables: Current achievements and future challenges. Acta. Hort. 2003;628:703-713.

19.   Coelho AR, Celli MG, Ono EYS, Wosiacki G, Hoffman FL, Pagocca FC, Hirooka EY. *Penicillium expansum* versus antagonist yeasts and patulin degradation in vitro.Braz. Arch. Biol. Techol.2007;50(4): 1-11.

20.   Pimenta RS, Silva FL, Silva JFM, Morais P B, Rosa DCA, Correa A .Biological control of *Penicillium italicum, P.digitatum* and *P. expensum* by the predacious yeast *Saccharomyces schoenii* on oranges. Braz. J. Microbiol. 2008;39(1):1-10.

21.   Miceli MH, Díaz JA, Lee SA. Emerging opportunistic yeast infections. Lancet Infect Dis. 2011;11:142–51.

22.   Desnos-Ollivier M, Ragon M, Dromer F. *Debaryomyces hansenii* (*Candida famata*), a Rare Human Fungal Pathogen Often Misidentified as *Pichia guilliermondii* (*Candida guilliermondii*) J Clin Microbiol. 2008;46(10):3237–3242.

23.   Milner RJ, Picard J, Tustin R. Chronic episodic diarrhoea associated with apparent intestinal colonisation by the yeasts *Saccharomyces cerevisiae* and *Candida famata* in a German Shepherd dog. Journal of the South African Veterinary Association. 1997; 68(4):147–149.

24.   Riley CB, Yovich JV, Robertson JP, O'Hara FL. Fungal arthritis due to infection by *Candida famata*. Australian Veterinary Journal. 1992;69:65–66 .

25.   Harmal NS, Khodavandi AMA, Alshawsh MA, Jamal F, Sekawi Z, Peng NK, Chong PP. Identification and differentiation of *Candida* species using specific polymerase chain reaction (PCR) amplification of the phospholipase B gene. African Journal of Microbiology Research. 2013;7(20):2159-2166.

26. White PL, Perry MD, Barnes RA. An update on the molecular diagnosis of invasive fungal disease. FEMS. Microbiol. Lett. 2009;296:1-10.

27. Opara CN, Okolie PI, Uzochukwu SVA. Diversity of bacterial community of African oil bean seeds (*Pentaciethra macrophylla* Benth) by comparison of 16 S rRNA gene fragments. British Biotechnology Journal. 2013;3(2):213-220.

10

Activity of β-Amylase in Some Fungi Strains Isolated from Forest Soil in South-Western Nigeria

O. A. Oseni[1*] and M. M. Ekperigin[2]

[1]Department of Medical Biochemistry, College of Medicine, Ekiti State University, Ado-Ekiti, Nigeria.
[2]Department of Biochemistry, Federal University of Technology, Akure, Ondo State, Nigeria.

Authors' contributions

This work was carried out in collaboration between all authors. Author MME designed the study, author OAO performed the statistical analysis, wrote the protocol and the first draft of the manuscript, managed the analyses of the study and the literature searches. All authors read and approved the final manuscript.

ABSTRACT

Aims: The aim of this study was to isolate fungi species from Omo natural forest soil in Ogun State of Nigeria and study the amylases from the fungi species which are digestive enzymes that hydrolyze glycosidic bonds in starch to glucose, maltose, maltotriose and dextrin; and in particular determine the activities of β-amylase from the forest soil.
Study Design: Nine different species of fungi were isolated from the Omo natural forest reserve soil *(Gonatobotrrys simplex, Aspergillus niger, Spiromyces minutus, Aspergillus flavus, Articulospora inflata, Botrytis cenera, Penicillium italicum, Aspergillus fumigatus and Aspergillus flavus).* Four species of the fungi *(Aspergillus flavus, Aspergillus niger, Aspergillus fumigatus and Penicillum italicum)* exhibited amylolytic activities maximally were obtained and screened for the production of beta-amylase (1,4-α-D-glucan maltohydrolase) for five days in liquid medium using 2% starch as carbon source. All the strains of fungi produced β-amylase optimally within the first 24 hours with progressive decreased production as the days gone by.
Place and Duration of Study: Department of Biochemistry, Federal University of Technology, Akure, Ondo State, Nigeria, between February 2010 and March 2011.
Methodology: We isolated many fungi species from forest reserve soil, four species

---

Corresponding author: Email: ooseni2003@yahoo.com

(*Aspergillus flavus*, *Aspergillus niger*, *Aspergillus fumigantus* and *Penicillum itallicum*) were identified and assayed for β-amylase activity.

Results: All the organisms produced β-amylase activity favorably at 40°C; all were observed to be thermally stable at between 30°C and 40°C with optimal pH in alkaline medium between pH 7.00 and 9.00.

Conclusion: The results obtained in this study however showed that all the fungi strains are promising sources of β-amylase for potential applications in food and pharmaceutical industries and for biotechnological and industrial applications.

Keywords: Forest-soil; fungi-strains; beta-amylase; optimum-activity; industry-application.

1. INTRODUCTION

The living organisms in the soil in terms of animals and plants are responsible for humus formation which is the product of degradation and synthesis in the soil. These organisms engineer myriad of biochemical changes as decay takes place [1], they also physically churn the soil and help stabilize soil structure. A vast number of organisms live in the soil [2]; so great are micro floral numbers that they dominate the biomass in spite of the minute size of each organism [3]. The influence of fungi is by no means entirely understood, but it is known that they play a very important part in the transformation of the soil constituents [4,5,6]. Over 690 species have been identified, representing 170 genera.

In the ability to decompose organic residues, fungi are the most versatile and perhaps the most persistent of any other group. Cellulose, starch, gums, lignin, as well as the most easily affected proteins and sugars, readily succumb to their attack [4]. In affecting the processes of humus formation and aggregate stabilization, molds are more important than bacteria. They are especially active in acid forest soils but play a more significant role than is generally recognized in all soils. Moreover, fungi function more economically than bacteria in that they transform into their tissues a larger proportion of the carbon and nitrogen of the compounds attacked and give off as by-products less carbon dioxide and ammonium. Nevertheless, soil fertility depends in no small degree on molds since they keep the decomposition process going after bacteria and actinomycetes have essentially ceased to function [4].

Enzymes responsible for the breakdown of starch are widely distributed in nature. Among these are the amylases, which act on starch, glycogen and derived polysaccharides to hydrolyze the α - 1, 4- glycosidic linkages. The amylase may thus be divided into three groups: the α - amylases (endoamylases), β-amylases (exo-amylases) and glucoamylases. The substrate and culture media component greatly influence the nature of amylase enzyme produced [7]. In starch processing industries, immobilized cells were used to optimally exploit the amylase producing machinery of the cells of which the β - amylase-producing cells are employed for bioconversion of starch to maltose [8]. Nowadays amylase from these sources is vastly used in amylase production under extreme conditions of pH and temperature. Amylases are of great importance in biochemical processes involving starch hydrolysis. The spectrum of application has gone beyond foods, beverages and pharmaceuticals to other fields such as medical and analytical chemistry [9]. In the present day biotechnology, approximately 25% of the enzyme market is dominated by amylases [10]. Therefore the present investigation therefore dealt with isolation and characterization of fungi species from soil samples collected from a forest reserve using physiological and biochemical features and hence determination of amylolytic activity and some kinetics of the β-amylase from these fungi.

2. MATERIALS AND METHODS

2.1 Isolation of Fungi

Amylolytic fungi used in this study were isolated from soil of Omo natural forest reserve in Ogun State of Nigeria using the method reported by Kader and Omar [5]. 1.0g of the freshly collected soil was mixed with 9.0mL of sterile distilled water in sterile test tube. The samples were serially diluted. 0.5mL of 10^{-3} diluents was pipetted, poured and dispersed by swirling on potato dextrose agar (PDA) and incubated at 30°C for five days. The isolates were sub cultured until single pure isolate was obtained.

2.2 Identification of Fungi Species

Characterization method employed for the fungal isolates were made by both the inspection of colonial features, cellular characteristics at x 100 and x 40 microscopic magnification. Identification was done by employing the method of [11] and conventional techniques of isolating individual microorganisms and allowing them to grow and produce colonies.

2.3 Preparation of Enzyme Solution

With the aid of a sterile cork borer, a 5mm disk from the advancing edge of a 4 day old fungal isolates were separately inoculated into the cultivation medium (1.0 K_2HPO_4; 0.5 $MgSO_4.7H_2O$; 2.0 $NaNO_3$; 0.001 $FeSO_4.7H_2O$; 0.5 KCl) g/L). The culture medium was adjusted to pH 7.0 before sterilization, 2% (w/v) pure soluble starch as carbon source was used for β amylases. Incubation was carried out at 30°C for 5 days on a shaker incubator operated at 150rpm.

2.4 Microbial Enzymes Assays

Enzyme solution (0.5mL) was added to 0.5ml of substrate (1% soluble starch was prepared in 0.016M sodium acetate of pH 4.8) and incubated at 25°C for 3 minutes. 1mL of 3, 5-DNSA (color reagent) was added. The mixture was then heated in water bath set at 100°C for 5 minutes after which, the mixture was cooled and 10mL of water was added. The color so formed was read in a colorimeter at 540nm against a blank containing buffer only. A calibration curve was made with maltose (0.3 – 3.0μmoles) to convert colorimeter readings to unit of activity [12].

One unit of β- amylase activity produces or release one micromole of β- maltose from soluble starch per min at 25°C and pH 4.8 in a reaction system containing 0.5mL of 1% soluble starch in 0.016M sodium acetate buffer of pH 4.8 and 0.5mL of crude enzyme solution.

3. RESULTS AND DISSCUSSION

Nine different species of fungi were isolated from the soil of the Omo natural forest reserve as seen from Table 1, Four species (*Aspergillus niger, Aspergillus flavus, Aspergillus fumigatus, Aspergillus flavus*) exhibited amylolytic activities maximally, therefore, β-amylase activity was monitored for five days in cultivating medium and some kinetic parameters studied.

Table 1. Strains of fungi isolated from the soil

Soil description	Fungi isolates
Omo natural forest soil	*Gonatobotrrys simplex,*
	Aspergillus niger
	Spiromyces minutus,
	Aspergillus flavus
	Articulospora inflata
	Botrytis cenera
	Penicillium italicum
	Aspergillus fumigatus
	Aspergillus flavus

The β-amylase activity of the fungal isolates was monitored for five days as shown in Fig. 1. The results revealed that all the organisms produced the enzyme with highest activities at first

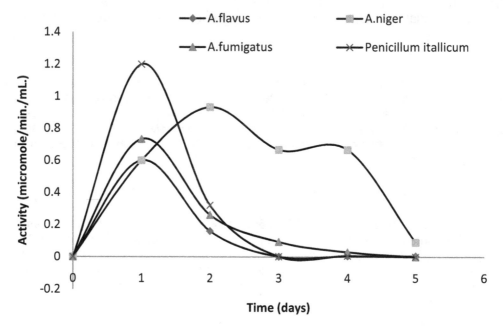

Fig. 1. β-amylase activity of different fungi isolates on daily basis

Day of culture showing *A. flavus,* with 0.6µmol./min/mL, *A. fumigatus* (0.733µmol./min/mL) and *P. italicum* 1.2µmol./min/mL, with exception of *A. niger* showing highest activity on second day of culture with 0.933µmol./min/mL. The results obtained in this work were in consonance with earlier observations of [13] that screened different species of *A. niger* for the synthesis of amylolytic enzymes using the submerged fermentation. Similarly [14] also obtained *A. flavus* with highest ability for amylolytic enzyme production among the isolated filamentous fungi from cereals, while [15] isolated a fungal strain, *Aspergillus tamari* from the soil having the ability to produce amylolytic enzymes, which formed both alpha amylase and glucoamylase in the mineral medium with 1.0% starch and maltose as carbon source respectively. The observation in this study also substantiate the earlier work of [16] where some fungi hydrolytic enzyme activities among which β-amylase were investigated from *Aspergillus niger* and *Penicillium italicum* isolated from natural forest and *Gmellina* plantation soils in Ondo State of Nigeria.

Similarly the 2 days maximum amylase production for *A. niger* in this study was in consonance with the observations of [17] for *A. ochraceus* and [18] where incubation period of between 1 and 2 days was observed to be the optimum time for all the fungi studied as increased incubation period decreased amylase activity. The reduced and subsequent diminished activity in the later phase of growth might be as a result of catabolite repression by glucose released from starch hydrolysis as earlier observed by [19] for *Humicola grisea,* [11] for *H. brevis,* but different from [20,21] for *Papulasporia thermophilia* in which the maximum amylase activity was recorded during the period of fungus autolysis.

The optimum temperature of 40^0C was observed for all the studied organisms with highest production of beta-amylase activity of *A. flavus,* *(*0.4μmol./min/mL), *A. niger* (0.667 μmol./min/mL), *A. fumigatus* (0.75μmol./min/mL) and *Penicillum itallicum* *(*0.883μmol./min/mL) as obtained from Fig. 2.

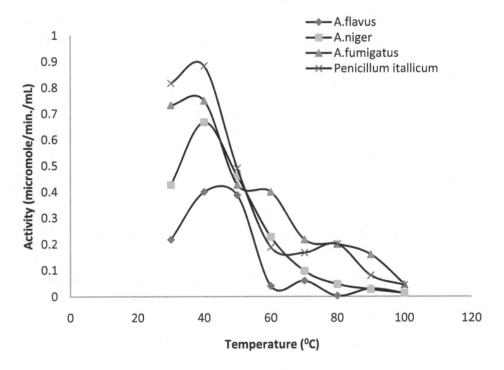

Fig. 2. Effect of temperature on β-amylase activity of different fungi isolates.

This observation however agreed with the works of [22,23,10,24] for studied fungi, but fail to agree with [25] who reported an optimum temperature of 55^0C for a fungus *Humicola grisea var. thermoidea* from Brazillian soil sample. Similarly the results of this study corroborates [26] in his investigation on production of microbial protease from selected soil fungal isolates where the optimum temperature was observed to fall between 30°C and 60°C but totally lost activities at between 80°C and 100°C for different fungi species.

The results obtained from this study also revealed that all the organisms were thermally stable and produced beta-amylase even at 80°C but lost activity completely at between 90°C and 100°C as seen from Fig. 3. However, this observation agreed with earlier work of [26] that microbial enzymes are thermally stable up till 80°C.

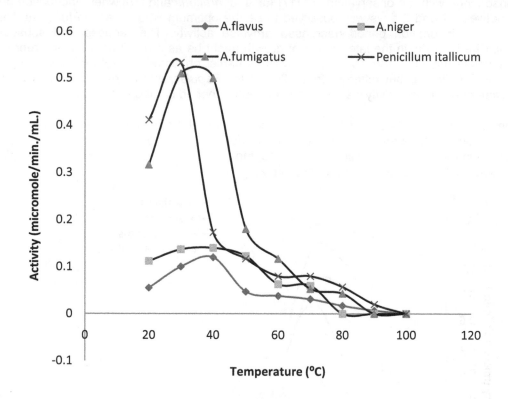

Fig. 3. Effect of Temperature on thermal stability of β- amylase activity of fungi isolates

The results of the study also showed that the organisms produced beta-amylase at both alkaline and acid range but the maximum pH recorded for the production of beta-amylase by all the organisms was at 8.5 as seen from Fig. 4, this observation agreed with [26,27] who obtained an optimum pH of between 8 and 9 for microbial enzymes activity from various fungal isolates though the activity of these enzymes were relatively high at pH between 3.0 and 9.0.

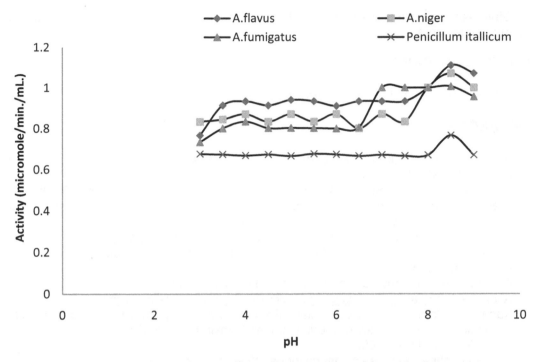

Fig. 4. Effect of pH on β-amylase activity of fungi isolates

However, various pH range has been reported for amylase production by different organisms; [28] reported pH range of 7.0-7.5 for maximum amylase production by *A. oryzae,* even though activity was observed at pH range of 5.0-10.0, while [29,30] reported a pH range of 6.0 and 8.0 for *A. fumigatus* as [31] reported a pH range of 3.0-7.0 for amylase production by *Aspergillus niger* though [32] reported amylase activity at pH range of 4.0-8.0, with maximum at pH 7.0 for *Talaromyces emersonii*. [17] also observed amylase production in the pH range of 4.0 to 6.0 for *Aspergillus ochraceus*. The differences observed however might be due to the source of isolation, the strains of the particular fungi and the type of culture medium used.

4. CONCLUSION

Conclusively, the results obtained in this investigation revealed that the studied fungi strains are promising sources of amylase for biotechnological and industrial applications, especially in medical and pharmaceutical uses, starch industry for ethanol production, production of high-fructose corn syrup, production of shorter chains of sugars called oligosaccharides and production of dishwashing and de-starching detergents and host of other useful products. Therefore efforts will be geared up in elucidating the molecular and protein sequence of β-Amylase from these microorganisms in this studied location.

ACKNOWLEDGEMENT

We appreciate the efforts of Dr. (Mrs.) Aborishade, Mrs. Toyin Ojo, of Department of Biology, and Mr. Fred Akharayi of Microbiology Department, Federal University of Technology, Akure,

Ondo State, Nigeria for the assistance rendered during the isolation and characterization of these microorganisms.

COMPETING INTERESTS

Authors have declared that no competing interests exist.

REFERENCES

1. Brady NC. The Nature and Properties of soils, 8th eds. Macmillan Publishing Co., Inc. New York. 1974;111–129.
2. Schmidt S, Martin A, Schadt C, Lipson D, Meyer A, Costello E, Nemergut D, Oline D, West A, Metchalf J Profound seasonal changes in microbial diversity and function in an Alpine Environment. 2002; Assessed May 2005.
    Available: http://www.colorado.edu/epob/EPOBprojects/schmidtlab/mo/nsf2002/nsf2002.htm
3. Russell EW. Soil conditions and Plant Growth 10th eds. Williams Clowes & Sons Ltd. London. 1977;150.
4. Divne C, Previous and current research. 1998; Assessed May 2003.
    Available: http://alpha2.bmc.uu.se/cici/private/research/Research.html.
5. Kader AJ, Omar O, Feng LS. Isolation of cellulolytic fungi from the Bario highlands, Sarawak. ASEAN Review of Biodiversity and Environmental Conservation (ARBEC). 1999; Assessed Dec. 2003.
    Available: http://www.arbec.com.my/pdf/art12sepoct99.pdf.
6. Piriyaprin S, Sunanthapongsuk V, Limtong P, Leaungvutiviroj C, Pasda N. Study on soil microbial biodervisity in rhizosphere of vetiver grass in degradating soil. 17thWCSS, 2002; pp14 – 21 Assessed July 2005.
    Available: http://www.sfst.org/proceedings/17wcss_CD/papers//1896.pdf.
7. Srivastava RAK, Baruah JN. Culture conditions for production of thermostable amylase by *Bacillus stearothermophilus* Applied and Environmental Microbiology. 1986;52(1):179-184.
8. Ray RR, Jana SC Nanda G. β- amylase production by immobilized cells of *Bacillus megaterium* B6. J. Basic microbiology. 1995;35:2:113-116.
9. Pandey A, Nigam P, Soccol CR, Soccol VT, Singh D, Mohan R. Advances in microbial amylases. Biotechnol. Appl. Biochem. 2000;31:135-152.
10. Rao VB, Sastri NVS, Subba Rao PV. Purification and characterization of a thermostable glucoamylase from the thermophilic fungus *Thermomyces lanuginosus*. Biochemial Journal. 1998;193:379–387.
11. Barnett EA, Hunter CL. The relation of extracellular amylase, mycelium, and time, in Some thermophilic and mesophilic Humicola species. Mycopathol, Mycol Appl. 1972;44:131–141.
12. Bernfeld P. Amylases α and β. Methods in Enzymology, (Colowick, S. P., and Kaplan, N. O.,eds.). Academic Press Inc., New York. 1955;149.
13. Rousset S, Schlich P. Amylase production in submerged culture using principal component analysis. J. Ferment & Bioeng. 1989;68:339-343.
14. AbouZeid AM. Production, purification and characterization of an extracellular alpha-amylase enzyme isolated from Aspergillus flavus. Microbios. 1997;89(358):55-66.
15. Moreira MT, Feijoo G, Sierra-Alvarez R, Lema J, Field JA. Re-evaluation of the manganese requirement for the biobleaching of kraft pulp by white rot fungi. Bioresource Technology. 1999;70:255-260.

16. Oseni OA, Ekperigin MM, Akindahunsi AA, Oboh G. Studies of physiochemical and microbial properties of soils from rainforest and plantation in Ondo state, Nigeria. African Journal of Agricultural Research. 2007;2(11):605-609.
Available: http://www.academicjournals.org/AJAR ISSN 1991- 637X © 2007 Academic Journals.

17. Nahas E, Waldemarin MM. Control of amylase production and growth characteristics of *Aspergillus ochraceus*. Rev Latinoam Microbiol. 2002;44(1):5-10.

18. Aiyer PV. Effect of C:N ratio on alpha amylase production by *Bacillus licheniformis*. African Journal of Biotechnology; 2004;3:519-522.

19. Adams PR, Deploey JJ. Amylase production by *Mucor miehei* and *M. pusillus*. Mycologia. 1976;68:934-938.

20. Adams PR. Amylase and growth characteristics of *Papulaspora thermophilia*. Mycopathologia. 1985;90:81-83.

21. Chapman ES, Evans E, Jacobelli MC Logan AA. The cellulolytic and amylolytic activity of *Papulaspora thermophila*. Mycologia. 1975;67:608-615.

22. Mishra RS, Maheshwari R. Amylases of the thermophilic fungus *Thermomyces lanuginosus*: their purification, properties, action on starch and response to heat. Journal of Bioscience. 1996;21:653–672.

23. Rao VB, Maheshwari R, Sastry NVS, Subba Rao PV. A thermostable glucoamylase from the thermophilic fungus *Thermomyces lanuginosus*. Current Science. 1979;48:113–115.

24. Taylor PM, Napier EJ Fleming ID. Some properties of glucoamylase produced by the thermophilic fungus *Humicola lanuginosa*. Carbohydrate Research. 1978;61:301–308.

25. Tosi LRO, Terenzi HF, Jorge JA. Purification and characterization of an extracellular glucoamylase from the thermophilic fungus Humicola grisea var. thermoidea. Can J Microbiol. 1993;39:846–852.

26. Oseni OA. Production of Microbial Protease from Selected Soil Fungal Isolates Nig J. Biotech. ISSN: 0189 17131 www.biotechsocietynigeria.org. 2011;23:28-34.

27. Pagare RS, Ramdasi AM, Khandelwal SR, Lokhande MO, Aglave BA. Production and enzyme activity of an extracellular protease from Aspergillus niger and Bacillus subtilis. International Journal of Biotechnology & Biochemistry. 2009;5(3):3-8

28. Kundu AK, Das S, Gupta TK. Influence of culture and nutritional conditions on the production of amylase by the submerged culture of *Aspergillus oryzae*. J. Ferment. Technol. 1973;51:142-150.

29. Absida VA. Some extracellular enzymes associated with tow tomato fruit spoilage molds. Mycopathologia. 1985;9:101-108.

30. Mahmoud ALE. Different factors affecting growth and amylase production by fungi inhabiting poultry feeds. J. Basic Mi crobiol. 1993;33:187-192

31. Fadel-M. Production of thermostable amylolytic enzymes by *Aspergillus niger* F-909 under solid state fermentation. Egyptian J. Microbiol. 2000, 35: 487-505.

32. Oso BA. Mycelial growth and amylase production by *Talaromyces emersonii*. Mycologia. 1979;71:521-529.

11

Biosorption of Lead by *Pleurotus florida* and *Trichoderma viride*

A. S. Arun Prasad[1*]**, G. Varatharaju**[1]**, C. Anushri**[1] **and S. Dhivyasree**[1]

[1]*Department of Biotechnology, Government College of Technology, Coimbatore -641013, India.*

Authors' contributions

This work was carried out in collaboration between all authors. Equal contribution was given by four authors. All authors read and approved the final manuscript.

ABSTRACT

Aims: The objective of the work is to remove Pb^{2+} by *Pleurotus florida* and *Trichoderma viride* in batch studies and to study the kinetics and adsorption isotherm of Pb^{2+} adsorption by fungal species and to determine the desorption performance by suitable desorbing agents.
Study Design: Experimental study.
Place and duration of the study: This work was carried out at Department of Biotechnology, Government College of Technology, Coimbatore, Tamil Nadu, and India for a period of five months.
Methodology: The polluted sample was collected from Valankulam lake, Coimbatore. The biomass of *Pleurotus florida* and *Trichoderma viride* were used as adsorbents. Atomic Absorption Spectrophotometer was used to quantify Pb^{2+} concentration. The optimum conditions of pH, adsorbent dose and contact time for biosorption were determined.
Results: Maximum adsorption of Pb^{2+} for *Trichoderma viride* and *Pleurotus florida* were observed at a pH of 6 and 7 respectively. The optimum quantities of adsorbent required for the removal of Pb^{2+} were 0.2g for both the organisms. Adsorption of Pb^{2+} was found to reach equilibrium in 1 h and 1.5 h for *Pleurotus florida and Trichoderma viride* respectively.
Conclusion: Hence, *Pleurotus florida* and *Tricoderma viride* are suitable adsorbents for the removal of Pb^{2+} from effluents. This methodology can be used for the removal of lead

*Corresponding author: Email: arunbiotech2006@yahoo.co.in;

from waste water before its disposal.

Keywords: Biosorption; toxic heavy metals; adsorbent; effluent.

1. INTRODUCTION

Heavy metals released by a number of industrial processes are major pollutants in marine, ground, industrial and even treated wastewaters. The most important characteristics of these metals are that they are non degradable and can accumulate in living tissue [1]. Therefore, for reducing lead mediated pollution in environment, Pb^{2+} should be removed from wastewater before its disposal [2]. The conventional methods for heavy metal removal from industrial effluents are precipitation, coagulation, ion exchange, cementation, electro dialysis, electro winning, electro coagulation and reverse osmosis [1,3,4].

Biosorption is a promising method for removal of toxic metal ions by living and dead microbial cells from aqueous solutions [5,6,7,8]. It can be efficiently used for the treatment of large volumes of effluents with low concentration of pollutants. So, the process does not depend on the viability of the biomass [9]. The applicability of fungi as biosorbent has some advantages due to their small size, ubiquity, ability to grow under controlled conditions and resilience to a wide range of environmental situations.

Biosorption is the binding and concentration of heavy metals from aqueous solutions (including very dilute concentrations) by certain types of inactive, dead and live microbial biomass [10,11]. Fungi, in common with other microbial groups, can accumulate metals from their external environment by means of physico-chemical and biological mechanisms [10]. Biosorption is a promising method for removal of toxic metal ions from waste water. Its advantage is especially in the treatment of large volumes of effluents with low concentration of pollutants. The major advantages of biosorption over conventional treatment methods include: Low cost, good efficiency, minimization of chemical and biological sludge, regeneration of biosorbent, and possibility of metal recovery [9]. The Pb^{2+} concentration in the sample was analyzed by Atomic Absorption spectrophotometer [10,12].

Fungi are chosen for biosorption because of their special physiology and adsorbing capacity. Chitin and chitosan present in fungi are well known metal ion adsorbers due to the presence of both carboxyl and amine groups.

2. MATERIALS AND METHODS

2.1 Media Used and Biomass Production

Oyster mushroom (*Pleurotus florida*) and *Trichoderma viride* were obtained from Tamil Nadu Agricultural University, Coimbatore. Potato Dextrose Agar and Molasses Yeast liquid medium were used for the maintenance and growth of *Trichoderma viride* and *Pleurotus florida* respectively [10,13]. 70ml of molasses medium was transferred to the 250 ml conical flask and autoclaved for 15 minutes. The mycelia disc of *Trichoderma viride* [14] was transferred to the flask and incubated for 10 days at 25°C. After three days, the growth of the fungus was observed in the form of pellicles, which increased in diameter on subsequent days. The mycelial mat was collected by filtering through muslin cloth after 10 days of growth and washed thoroughly with deionized water to remove the growth medium sticking on its surface [15].

2.2 Preparation of Standard and Adsorbents

Standard solutions of Pb^{2+} nitrate were prepared in the range of 0ppm to 20ppm using double distilled water. For regeneration studies, Hydrochloric acid and Sodium hydroxide of different concentrations of 0.01 M, 0.1M and 1M were used [10]. The obtained biomass of *Pleurotus florida* and *Trichoderma viride* were cleansed with distilled water and dried at 60°C for a period of 24h. The biomasses were ground into fine particles using electric grinder and mortar and pestle respectively. The particles were sieved using 300µm sieve and used for biosorption studies.

2.3 Pb^{2+} Analysis

The presence of Pb^{2+} in the sample was analysed by Atomic Absorption Spectro photometer (GBG 901, Australia). The samples for Pb^{2+} analysis were collected from Valankulam Lake (Coimbatore Corporation).

2.4 Batch Adsorption

The biosorption studies were conducted in batch process to evaluate the effect of pH, initial metal concentration and contact time at different concentrations of biomass on removal of Pb^{2+} ions. All biosorption experiments were carried out in 250ml of Erlenmeyer flasks in rotary incubator shakers at 140rpm and 30°C.

Metal uptake (q) can be determined by: [11,10]

$$Q = V \times (Ci - Ce) / S \qquad (1)$$

where,
Q (metal uptake, mg/ g) is the amount of metal ions adsorbed on the biosorbent.
V (ml) is the volume of metal containing solution in contact with the biosorbent.
Ci and Ce (mg /l) are the initial and equilibrium (residual) concentrations of metals in the solution, respectively and
S (g) is the amount of added biosorbent on dry basis.

Effect of pH on biosorption rate of the fungal biomass was investigated in the initial pH range from 3.0 to 10.0 at 30°C. The pH of the solution was adjusted using 1N HCl and 0.1N NaOH. 0.2g biomass of *Pleurotus florida* and *Trichoderma viride* were transferred to the flasks and the reaction mixture was shaken in rotary incubator shaker at 140 rpm for 3h. The initial concentrations of Pb^{2+} ions were varied from 4 mg/l to 20 mg/l at optimum pH, for 3h contact time. Similarly the contact time was varied from 30 min to 3h to determine optimum biosorption time at optimum pH. Effects of biosorbent dose were also investigated. Biomass dose were varied from 0.04g to 0.24g/100ml of 10mg/l Pb^{2+} solutions. Pseudo first order rate kinetics was applied and value of rate constant k_{ad} were derived from Lagergren plots.

2.5 Adsorption Isotherm

The adsorption isotherm is the initial experimental step to determine the feasibility of adsorption treatment. It is a batch equilibrium test, which provides data relating adsorbate per unit weight to the amount of adsorbate remaining in the solution. Adsorption data for a wide range of adsorbate concentrations are most conveniently described by various adsorption isotherms, namely Langmuir or Freundlich isotherm [15].

The Langmuir model can be described as [4,17,18]

$$q_e = Q_0 \, b \, C_e \, / \, 1 + bC_e \qquad (2)$$

where,
q_e is the uptake of metal ions per unit weight of the adsorbent.
Q_0 is the moles of solute sorbed per unit weight of adsorbent.
b is the constant relates the affinity between the biosorbents and biosorbate.
C_e equilibrium concentration of ions [15,16].

The constants, Q_0 and b are evaluated from the linear plot of the logarithmic equation.

$$1/q_e = 1/ \, Q_0 + 1/bQ_0 \times 1/C_e \qquad (3)$$

The Langmuir model is based on the assumption that maximum adsorption occurs, when a saturated monolayer of solute molecule is present on the adsorption surface. The energy of adsorption is constant and there is no migration of adsorbate molecule in the surface plane.

The Freundlich isotherm is of the form: [4,17]

$$q_e = k \, C_e 1/n \qquad (4)$$

The logarithmic form of the above equation is as follows:

$$\log q_e = \log k + 1/n \log C_e \qquad (5)$$

where,
q_e is the uptake of metal ions per unit weight of biosorption.
C_e is the equilibrium concentration of metal ions in solution.
k is the Freundlich constants denoting adsorption capacity.
n is the empirical constants, is a measure of adsorption intensity [15,16].

The value of k and 1/n were found by plotting the graph between $\log q_e$ and $\log C_e$. The value of log k is the intercept and value of 1/n is the slope of the plot. The value of k is determined from the antilog of the intercept value. A high value of 'k' and 'n' indicates high adsorption throughout the concentration range and vice-versa. A low value of 'n' indicates high adsorption at strong solute concentration [19].

3. RESULTS AND DISCUSSION

3.1 Results

Biosorption of Pb^{2+} by *Pleurotus florida* and *Trichoderma viride* are influenced by several factors like pH, initial Pb^{2+} ion concentration, adsorbent dose and contact time.

3.1.1 Effect of pH

The effect of pH on the biosorption of Pb^{2+} by *Pleurotus florida* (A) and *Trichoderma viride* (B) were studied with at pH 3.0 to 10.0. The results are presented in Fig (3.1) and Fig (3.2) respectively. The maximum biosorption of Pb^{2+} by A and B were observed at neutral and acidic pH respectively. It was found that the removal of Pb^{2+} was 100% with *Pleurotus florida* at pH 7.0 and 90% with *Trichoderma viride* at pH 6.0.

3.1.2 Effect of adsorbent dose

The influence of the amount of adsorbent dose on Pb^{2+} adsorption was studied at 30°C and pH 7.0 for A and pH 6.0 for B respectively, by varying the adsorbent dose from 0.04 g to 0.24 g, while keeping the volume and concentration of the metal solution constant. This is shown in Fig (3.3) & Fig (3.4). But it is apparent that the percentage removal of Pb^{2+} increases rapidly with increase in the dose due to greater availability of the biosorbent. Adsorption is maximum with 0.2 g of biosorbent, when the biosorption concentration was increased from 0.04 g to 0.2 g in 100ml,the Pb^{2+} ion concentration were decreased from 10 mg/g to 5 mg/g for *Pleurotus florida* and 7.5 mg/g to 4.5 mg/g for *Trichoderma viride*.

3.1.3 Effect of contact time

Maximum Pb^{2+} removal with fungal biosorbents was in initial periods of 60 min for *Pleurotus florida* and 90 min for *Trichoderma viride*, after which no significant removal of Pb^{2+} was observed with both the biosorbents. It is shown in Fig (3.5) & Fig (3.6). This rapid sorption stage indicates that surface sorption occur on the fungal cell surface. The kinetics of metal adsorption on the surface is usually rapid during initial period of time. In order to analyze whether sorption of Pb^{2+} follows pseudo first order reaction, kinetic experiments were carried out for a regular interval of 5 min for A and 10 min for B at 10 ppm of 0.2 g of adsorbent [15,1].

Lagergren Pseudo First Order Kinetics [20]

$$\log (q_e - q_t) = \log q_e - k_{ad} / 2.303 \times t \qquad (6)$$

where, k_{ad}(min) is the rate constant of adsorption; q_e and q_t are the amount of Pb^{2+} adsorbed(mg/g) at equilibrium and any time t min. This is shown in Fig (3.11) & Fig (3.12). The k_{ad} values are calculated from the graph plotted between $\log (q_e - q_t)$ and time t (min) [8].The straight lines and the value of R_2 confirm that the adsorption process follow first order rate kinetics in each case. The value of pseudo first order rate constant was found to be 2.23×10^{-2} for A and 3.385×10^{-2} for B.

3.1.4 Adsorption experiment

The Pb^{2+} adsorption followed Freundlich and Langmuir model. The value of Freundlich constant (k and n) and Langmuir constants (Q_0 and b) were evaluated from Fig (3.7) & Fig (3.8) for A and Fig (3.9) & Fig (3.10) for B. The high value of Langmuir constant Q_0 (12.195 mg/g) and b (0.058) with *Pleurotus florida* than *Trichoderma viride* Q_0 (10.989 mg/g) and b (0.057) indicated better adsorption capacity. Higher values of k and n, and lower values of b indicate better affinity of the biomass.

3.1.5 Regeneration studies

The applicability of fungal biomass for metal ion recovery from waste stream requires that, the biomass be regenerated efficiently. Thus the bound metal can be recovered in concentrated form and the biomass can be reused. Regeneration of adsorbed Pb^{2+} on to biomass was performed by taking different concentration of Hydrochloric acid and Sodium hydroxide at 0.01 M, 0.1 M, and 1 M of 100 ml solution. Metal adsorbed biomass was separated from equilibrium solution by means of centrifugation. It is found that the acidic solution desorption capacity was higher when compared to alkaline solution. For *Pleurotus florida* the order of desorption was 0.1 M HCl> 0.01 MHCl> 1 M HCl> 0.1 M NaOH > 0.01 M NaOH > 1 M NaOH. At 1 M NaOH there is no desorption. In *Trichoderma viride* the order of desorption was 1 M HCl> 0.01 M HCl >0.1 M HCl > 1 M NaOH > 0.1 M NaOH > 0.01 M NaOH. The results are shown in Fig (3.13) & Fig (3.14).

3.2 Discussion

The effect of pH on biosorption is due to interaction of Pb^{2+} cations with the carboxyl groups of chitosan or chitin present on the fungal surface. Effect of adsorbent dose is attributed to reduction of total area of biosorbent due to aggregation during biosorption. The rapid sorption stage indicates that biosorption occurs only on the fungal surface. The extent of adsorption efficiency increases with time and attains equilibrium at 90 minutes for both the adsorbents.

3.3 Tables and Figures

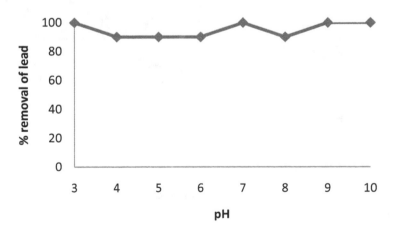

Fig. 3.1 Effect of pH on biosorption of Pb^{2+} by *Pleurotus florida*

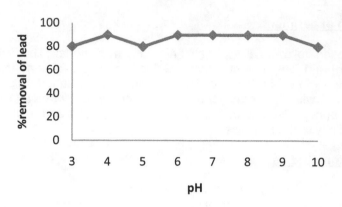

Fig. 3.2 Effect of pH on biosorption of Pb^{2+} by _Trichoderma viride_

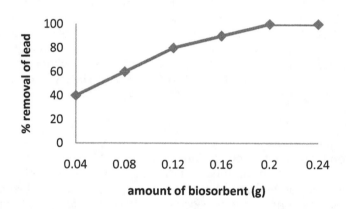

Fig. 3.3 Effect of biosorbent dose on Pb^{2+} biosorption by
Pleurotus florida

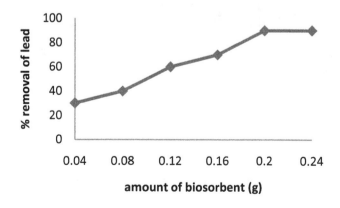

Fig. 3.4 Effect of biosorbent dose on Pb^{2+} biosorption by _Trichoderma viride_

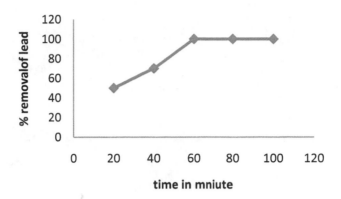

Fig. 3.5 Effect of time on Pb²⁺ biosorption by *Pleurotus florida*

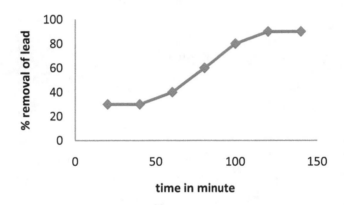

Fig. 3.6 Effect of time on Pb²⁺ biosorption by *Tricoderma viride*

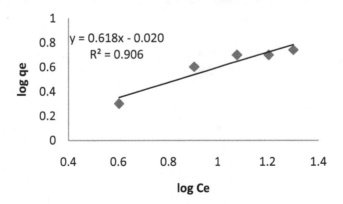

Fig. 3.7 Freundlich isotherm for Pb²⁺ removal by
Pleurotus florida

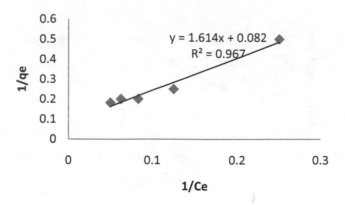

Fig. 3.8 Langmuir isotherm for Pb^{2+} removal by *Pleurotus florida*

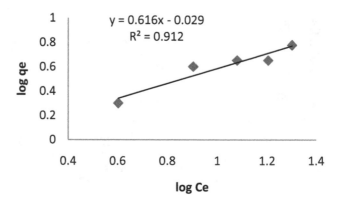

Fig. 3.9 Freundlich isotherm for Pb^{2+} removal by *Trichoderma viride*

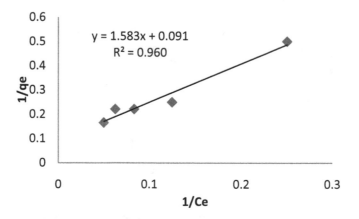

Fig. 3.10 Laungmuir isotherm for Pb^{2+} removal by *Trichoderma viride*

Table 3.1 Langmuir isotherm and Freundlich isotherm constants, coefficient of correlation

Biosorbent	Freundlich constants			Langmuir constants		
	K(mg/g)	N	r^2	Q_0(mg/g)	b(l/mg)	r^2
Pleurotus florida	0.954	1.610	0.906	12.195	0.058	0.967
Trichoderma viride	0.935	1.623	0.912	10.989	0.057	0.960

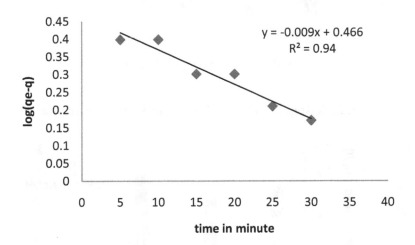

Fig. 3.11 Lagergren Pseudo first order kinetics for Pb^{2+} removal by
Trichoderma viride

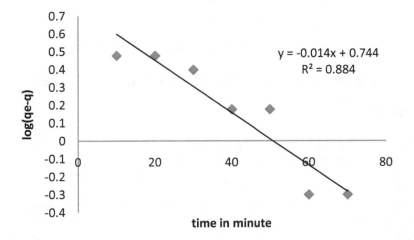

Fig. 3.12 Lagergren Pseudo first order kinetics for Pb^{2+} removal by
Pleurotus florida

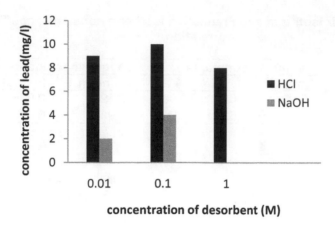

Fig. 3.13 Regeneration of Pb^{2+} from *Pleurotus florida*

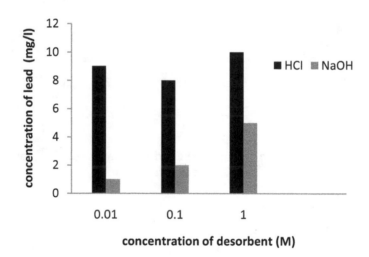

Fig. 3.14 Regeneration of Pb^{2+} form *Trichoderma viride*

4. CONCLUSION

Pleurotus florida and *Trichoderma viride* are suitable for the removal of Pb^{2+} from polluted sample. The adsorption is strongly dependent on pH and adsorbent dose. A maximum of about 90% Pb^{2+} removal was achieved at pH 6 for *Trichoderma viride* and 100% at pH 7 for *Pleurotus florida*. Removal of Pb^{2+} by batch studies showed that 0.2 g of *Pleurotus florida* and *Trichoderma viride* were the optimum quantities required for the removal of Pb^{2+} from 100mL of 10 mg/l Pb^{2+} nitrate solution. Adsorption of Pb^{2+} is fairly rapid in first 20 min and increased slowly to reach equilibrium in 1 h for *Pleurotus florida* and 1.5 h for *Trichoderma viride*. The regression analysis of equilibrium data fitted more to Langmuir adsorption isotherm than Freundlich isotherm. The adsorption of Pb^{2+} follows pseudo first order kinetics. Regeneration study revealed that exhausted biosorbent may be regenerated and used for nearly 6 cycles without significant loss of sorption capacity indicating that fungal biosorbent possesses good attritional characteristics. Over the last few decades, the huge increase in the use of heavy metals has resulted in an increased flux of metallic substances in aquatic environment. So, taking into consideration the health issues of living organisms it is mandatory to remove Pb^{2+} from waste water.

ACKNOWLEDGEMENTS

The authors are cordially grateful to The Principal and Head of the Department, Government College of Technology for being a constant source of encouragement and support.

The authors wish to express our gratitude to Dr. S. Nakeeran and Dr. A. S. Krishnamoorthy, Department of Plant Pathology, Tamil Nadu Agricultural University, Coimbatore for their valuable suggestions, persistent encouragement and keen involvement throughout this work which were of immense help in successfully completing the work.

Authors also thank Dr. Issac Soloman Jebamani and Mr. B. K. Soundaraj who provided us instrumental support.

COMPETING INTERESTS

Authors have declared that no competing interests exist.

REFERENCES

1. El-Ashtoukhy et al. Removal of Pb^{2+}(II) and copper (II) from aqueous solution using pomegranate peel as a new adsorbent. Desalination. 2008;223:162–173.
2. Zhan XM, Zhao X. Mechanism of Pb^{2+} adsorption from aqueous solutions using an adsorbent synthesized from natural condensed tannin. Water Res. 2003;37(16):3905-12.
3. Elias RW, Gulson B. Overview of Pb^{2+} remediation effectiveness. The Science of the Total Environment. 2003;303:1–13.
4. Favero N, Costa P, Massimino ML. In vitro uptake of cadmium by basidiomycete *Pleurotus ostreatus*. Biotechnol Lett. 1991;10:701–704.
5. Kapoor A, Viraraghavan T. Heavy metal biosorption sites in *Aspergillus niger*. Bioresource Technology. 1997;61:221-227.
6. Kapoor A, Viraraghavan T. Biosorption of heavy metals on *Aspergillus niger*: Effect of pretreatment. Bioresource Technolog. 1998;63:109-113.
7. Pighi PL, Pumpel T, Schinner F . Selective accumulation of silver by fungi. Biotechnol. 1989;11:275-280.
8. Siegel SM, Galun M, Siegel BZ . Filamentous fungi as metal biosorbents: a review. Water, Air Soil Poll. 1990;53:335-344.
9. Das N., Vimala R, Karthika P. Biosorption of Heavy metals- An review. Indian journals of biotechnology. 2008;7:159-169.
10. Abuk A, Ulhan S, Fuluk U, Caliskan F. Pb^{2+} biosorption by pretreated fungal biomass. Turk J Biol. 2005;29:23-28.
11. Das N, Charumathi D, Vimala R. Effect of pretreatment on Cd^{2+} biosorption by mycelial biomass of *Pleurotus florida*. Afr. J. Biotechnol. 2007;6(22):2555-2558.
12. Volesky B, Holant ZR. Biosorption of Heavy Metals. Biotechnol. 2009;11:235-250.
13. Hayyan Ismaeil Al-Taweil, Mohammad Bin Osman, Aidil AH, Wan Mohtar Wan Yussof. Optimizing of *Trichoderma viride* cultivation in submerged state fermentation. Am. J. Applied Sci. 2009;6:1277-1281.

14. Ujor VC, Monti M, Peiris DG, Clements MO, Hedger JN. The mycelial response of the white-rot fungus, Schizophyllum commune to the biocontrol agent, *Trichoderma viride*. Fungal Biol. 2012;116(2):332-341.

15. Bai SR, Abraham TE. Biosorption of Cr(VI) from an aqueous solution by *Rhizopus nigricans*. Biores Tech. 2009;79:73.

16. Bishnoi et al. Biosorption of Cr(VI) with *Tricoderma viride* immobilized fungal biomass and cell free Ca-alginate beads. Indian Journals of Experimental Biology. 2007;45:657-664.

17. Gabriel J, Mokrejs M, Bily J, Rychlovsky P. Accumulation of heavy metals by some wood-rotting fungi. Folia Microbiol. 1994;39:115–118.

18. Namasivayam C, Ranganathan K. Removal of Fe (II) by waste fe(III)/Cr (III) hydroxide from aqueous soln and electroplating industry waste water. Indian Journal of Chemical Technology. 1995;32:351.

19. Regine HSF. Vieira, Boya Volesky. Biosorption: a solution to pollution? Internatl Microbiol. 2000;3:17–24.

20. Narsi R Bishnoi, Garima. Fungus-An alternative for bioremediation of heavy metals containing waste water: A review. Journal of Scientific & Industrial Research. 2005;64:93-100.

12

Effects of Utilization of Crushed, Boiled and Fermented Roselle Seeds *(Hibiscus sabdariffa)* on the Performance of Broiler Chickens

Maikano Mohammed Ari[1*], Danlami Moses Ogah[1],
Idris Danladi Hassan[2], Ibrahim Suleiman Musa-Azara[2],
Nuhu Dalami Yusuf[2] and Samuel Emmanuel Alu[1]

[1]*Department of Animal Science, Faculty of Agriculture, Nasarawa State University Keffi, Shabu- Lafia Campus, Nasarawa State, Nigeria.*
[2]*Department of Animal Production, College of Agriculture Lafia, Nasarawa State, Nigeria.*

Authors' contributions

All authors were involved in the design of the study, statistical analysis, wrote the protocol, and wrote the draft of the manuscript. The corresponding author MMA is the lead Researcher.

ABSTRACT

Aim of Study: The study examines the effects of utilization of crushed (CRRS), boiled (HTRS) and fermented (FRS) Roselle seeds *(Hibiscus sabdariffa)* on the performance of broiler chickens

Study Design: A total of 135 Anak day-old broiler chicks were randomly assigned to three (3) experimental groups of three (3) replicates using completely randomized design, data collected were subjected to ANOVA using SPSS and Likert scaling technique.

Place and Duration of Study: Livestock Complex, College of Agriculture, Lafia, Nasarawa state, Nigeria: February 2012 to April 2012.

Methodology: The effects of inclusion of differently processed Roselle seeds on performance traits of experimental birds were evaluated through feeding trials (1- 28 d) and (29- 50 d) at starter and finisher phases. Dietary treatments were as follows: D1, D2 and D3 representing Crushing of Raw Rosselle Seeds (CRRS); Hydrothermally Processed Rosselle Seeds (HTRS) and Fermented Rosselle Seeds (FRS) base diets.

Results: No significant ($P=0.05$) difference in the following parameters: initial weight, feed

Corresponding author: Email: arimaikano@yahoo.com

intake, FCR and survival percentage in the starter phase while the finisher phase significantly (*P*=0.05) differ only the performance index. Overall scoring of performance parameters showed those birds in D3 group were better than D2 and D1 in that order **Conclusion:** Roselle seeds inclusion in broiler diets provides effective mechanism for the improvement in performance traits of broilers and fermentation of rosselle *(Hibiscus sabdarif)* is best processing method.

Keywords: Broilers; Rosselle seeds; Processing and Performance traits.

1. INTRODUCTION

Alternative plant protein sources and their nutritional potentials in the diets of monogastrics and human in developing countries like Nigeria have been highlighted [1, 2, 3 and 4]. *Hibiscus Sabdariffa* is fast gaining prominence as one of the replacement of rich plant proteins in human, livestock and fisheries nutrition on account of cost and nutritional composition [5]. Roselle seed meal is presently sold in some Nigerian markets at one third of the cost of soybean meal, thus Roselle seed meal represent attractive replacements for soybean meal in the diets of broilers from the standpoint of economics, availability and nutritional value.

Other nutritional benefits of Roselle seeds documented include the bioavailability of its nutrients particularly digestible protein (DP) and digestible energy (DE) [6,7]. However, these benefits are dependent on many factors, these include among others, sources and species of hibiscus and the processing method used before inclusion into the diets.

Therefore, the objectives of this study is to evaluate the effects of inclusion of raw, thermally processed and fermented Roselle seeds in the diets of broilers and its effects on performance traits.

2. MATERIALS AND METHODS

2.1 Experimental Site

This study was conducted at the Livestock Complex of College of Agriculture, Doma Road, Lafia which is located between latitude 8° and 9° North and longitude 80^0 and 90^0 East. The minimum temperature is $21.9^\circ C$ and maximum temperature of $37.6^0 C$ between January to June and the average annual rainfall is 823mm. The test ingredients were processed at both the Livestock Complex and the Nutrition Laboratory of the College, while the final feed was compounded at the feed mill unit of the complex.

2.1.1 Rosselle seed collection, processing and diet preparation

Rosselle seeds *(Hibiscus Sabdariffa)* was procured from a local market in Langtang South of Plateau State, Nigeria. The collected seeds were cleaned by winnowing and hand picking of stones and debris. The raw Roselle seeds were subjected to three processing methods viz: crushing of raw Roselle seeds (T1), hydrothermal (T2) and fermentation (T3). Each of these processing methods of Roselle seed served as test ingredient and was used as a replacement of soyabean meal in the broiler diets; these represent experimental treatment groups. The different processing methods of Roselle seed are described as thus:

2.1.1.1 Crushing of raw rosselle seeds (CRRS) - (T1)

Roselle seeds were cleaned by removing dust, stones and plant debris. The seeds were milled using a laboratory scale hammer miller and sieved through a 30mm mesh screen according to the methods described by [5]. The milled and bagged Roselle seed represents experimental treatment (1) (CRRS).

2.1.1.2 Hydrothermally processed rosselle seeds (HTRS) –T2

The method adopted by [8] was used. The cleaned seeds were poured into aluminum tower pot containing 50 litres of clean water in a batch of 50 Kg. The Roselle seeds sample was allowed to boil at $100^{o}C$ for 30 minutes before cooling by spreading on jute bags until stable weight was attained at room temperature. The cooked, milled and bagged Roselle seed represents experimental treatment (2) (HTRS).

2.1.1.3 Fermented Rosselle Seeds (FRS) –T3

The raw roselle seeds were sorted to ensure cleaned grains. The cleaned Roselle seeds were poured into a drum of 50 litres of boiling water per batch of 50kg Roselle seeds and allowed to boil at $100^{o}C$ for 30 minutes according to the method described by [2]. The boiled grains were drained, cooled to room temperature and placed in a leaf-lined basket covered with further leaves and kept for 48 hours. The products were sum dried and milled. The Fermented, milled and bagged Roselle seed represents experimental treatment (3) (FRS).

2.2 Experimental Treatment

A total of 135 Anak day-old broiler chicks were randomly divided into three (3) experimental groups of three replicates each. Dietary treatments were as follows: D1, D2 and D3 which had: Crushing of Raw Rosselle Seeds (CRRS); Hydrothermally Processed Rosselle Seeds (HTRS) and Fermented Rosselle Seeds (FRS) as replacement of soyabeans at both starter and finisher phases using completely randomized design. The incorporation of test ingredients served as the main source of variations. The starter diets were fed for four (4) weeks (1- 28 d) brooding phase and the finisher diets were fed for three (3) weeks (29- 50 d). The experimental feeds were formulated using a least cost feed formulation software *Feedwin*.

2.3 Data Collection

2.3.1 Performance parameters

The following parameters were measured and computed from the data generated from daily and weekly recordings during the feeding trials: feed intake, body weight, bodyweight gain, feed conversion ratio (feed: gain), mortality, survival percentages and performance index according to the methods adopted by [9].

2.3.2 Chemical analysis

The chemical composition of each of the processed rosselle seed samples and experimental diets were determined following standard methods [10]. Crude protein (N*6.25) was determined by the Kjeldahl method after acid digestion (Gerhardt, Königswinter, Germany).

Crude fat analysis without acid hydrolysis was determined by the ether-extraction method using a Soxtec System (Gerhardt, Königswinter, Germany). Moisture was determined by oven drying at 105°C until a constant weight was achieved. Ash content was estimated by incinerating the samples in a muffle furnace at 600 °C for 6 h. Total Carbohydrates (Nitrogen-free extract) was determined by difference and calculated thus: 100% - %(CP+ Ash + Crude Fat + Moisture).

2.4 Statistics

Data collected were subjected to One-way Analysis of Variance (ANOVA), means were separated (P=0.05) where there were significant differences using Duncan's Multiple Range Test [11] using SPSS 16.0 [12]. Qualitative evaluation techniques (Likert scaling) according to the methods described by [13] was used to draw conclusions on overall parameters measured.

3. RESULTS AND DISCUSSION

The chemical composition of the crushed, boiled and fermented Roselle seeds is presented in Table 1. Dry matter (DM) values ranged from 90.40 to 91.39% respectively for hydrothermal and crushed seeds while crude protein (CP) ranged from 19.54% to 23.43% for crushed and fermented seeds respectively. Crude fibre (CF) on the other hand had values ranging from 3.30% to 4.52% for fermented and crushed seeds respectively. The highest value of ether extract (EE) was obtained in fermented rosselle (6.58%) while the least (5.70%) was obtained in crushed seeds. Total ash ranged from 5.39% to 6.94% while total carbohydrates expressed as nitrogen free extract (NFE) ranged from 50.43 to 55.36%. The highest calcium (Ca) and phosphorous (P) values were 1.12 and 0.56% in hydrothermal and fermented Roselle seeds respectively. These findings are similar to the chemical composition of Roselle seeds reported by [14], [15] and [16].

Table 1. Effect of processing on the Chemical composition of Rosselle (*Hibiscus sabdariffa*) seeds

Methods of Processing	Chemical composition (%)							
	Dry Matter	Crude protein	Crude Fibre	Ether Extract	Total ash	NFE	Ca	P
Crushing of raw seeds	91.39	19.54	4.52	5.70	6.27	55.36	0.71	0.38
Hydrothermal	90.40	21.84	3.60	5.85	5.39	53.72	1.12	0.56
Fermentation	90.68	23.43	3.30	6.58	6.94	50.43	0.30	0.56

The composition of the experimental diets is presented in Table 2. The experimental diets are within recommended range [17]. The variations in the nutrient composition of the experimental diets at both starter and finisher phases were also within recommended range for birds in the tropics as reported by [18].

The performance parameters of broilers fed crushed, boiled and fermented Roselle seeds (*Hibiscus sabdariffa*) based diets are presented in Table 3. Effects of different processing methods of Roselle seeds did not differ significantly (P=0.05) in the following parameters: initial weight, weight gain, FCR and survival percentage in the starter phase while the finisher phase significantly (P=0.05) differ only in the performance index. The best values for body weight gain were broilers fed Diets 2 and 3 (536.36g and 990.38g) at the starter and finisher phases respectively. Diet 3 group presented higher feed intake values of 624.39g and 1272.67g in both the starter and finisher phase of the feeding trials. However, the best values recorded for FCR (1.18 and 1.41) in the starter and finisher phases were for diet 2 group. The survival percentage and performance index values were better with D3 group (94.67% and 4536.36) respectively in the starter phase while in the finisher phase D2 group presented better values of 96.5% and 8550.75 respectively for survival percentage and performance index.

Table 2. Composition of Experimental Diets

	Starter phase			Finisher phase		
	D1	D2	D3	D1	D2	D3
Maize	27.00	27.00	27.00	34.00	34.00	34.00
Maize Bran	3.25	4.00	4.00	3.25	3.25	3.25
Cassava	14.50	14.50	14.30	14.45	14.45	14.45
Soya toasted	18.55	17.80	18.00	13.00	13.00	13.00
CRRS	25.00	-	-	23.00	-	-
HTRS	-	25.00	-	-	23.00	-
FRS	-	-	25.00	-	-	23.00
Blood Meal	2.00	2.00	2.00	3.50	3.50	3.50
Fish Meal	5.00	5.00	5.00	3.00	3.00	3.00
Bone Meal	3.50	3.50	3.50	3.50	3.50	3.50
Palm Oil	0.50	0.50	0.50	1.60	1.60	1.60
L-Lysine	0.10	0.10	0.10	0.10	0.10	0.10
DL-Methionine	0.10	0.10	0.10	0.10	0.10	0.10
Salt	0.25	0.25	0.25	0.25	0.25	0.25
Premix*	0.25	0.25	0.25	0.25	0.25	0.25
Total	100	100	100	100	100	100
**Calculated						
ME/Kcal/kg	3126.85	3112..95	3104.73	3193.08	3185.74	3177.18
CP%	21.15	21.50	21.97	19.37	19.90	20.26
Determined analysis						
D M (%)	92. 73	92. 78	92. 69	92.95	93.43	92.83
CP (%)	21.05	21.33	21.73	19.83	19.95	20.17
CF (%)	4.82	4. 38	4.36	4.71	4.34	4.39
EE (%)	7.71	7.64	8.02	8.62	8.42	8.84
T Ash (%)	14.67	15.19	15.23	10.31	9.96	10.26
Ca	1.10	1.15	1.06	0.98	1.06	0.97
P	0.77	0.79	0.79	0.68	0.71	0.68
NFE (%)	43.93	43.04	34.08	49.48	50.76	49.17

*Premix to provide the following per KG of diet: Vitamin A, 9,000 IU; Vitamin D3, 2,000,IU; vitamin E, 18 IU; vitamin B1, 1.8 mg; vitamin B2, 6.6 mg B2,; vitamin B3, 10 mg; vitamin B5, 30 mg; vitaminB6, 3.0 mg; vitamin B9, 1 mg; vitamin B12, 1.5 mg; vitamin K3, 2 mg; vitamin H2, 0.01 mg; folic acid, 0.21 mg;nicotinic acid, 0.65 mg; biotin, 0.14 mg; Choline chloride, 500 mg; Fe, 50 mg; Mn, 100 mg; Cu, 10 mg; Zn, 85 mg;l, 1 mg; Se, 0.2 mg.

** Calculated using feedwin software

***DM (Dry Matter); CP (Crude Protein); CF (Crude Fibre); EE (Ether Extract); T Ash (Total Ash); Ca (calcium); P (phosphorous); NFE (Nitrogen Free Extract)

****Crushed Raw Rosselle Seeds (CRRS); Hydrothermally Processed Rosselle Seeds (HTRS); Fermented Rosselle Seeds (FRS)

Table 3. Effects of utilization of crushed, boiled and fermented roselle seeds (*Hibiscus sabdariffa*) on the performance of broiler chickens

Parameters	D1	D2	D3	SEM	D1	D2	D3	SEM
	Starter diets				**Finisher diets**			
Initial weight(g)	40.70[a]	41.32[a]	41.16[a]	±0.27	561.88[a]	580.66[a]	655.75[a]	±33.55
Feed intake(g)	544.33[b]	545.32[b]	624.39[a]	±16.78	1184.39[b]	1139.79[c]	1272.67[a]	±24.70
Weight gain(g)	408.44[a]	486.12[a]	536.36[a]	±45.60	812.66[a]	940.44[a]	990.38[a]	±136.25
FCR	1.34[a]	1.18[a]	1.22[a]	±0.10	1.54[a]	1.41[a]	1.44[a]	±0.20
Survival percentage (%)	91.00[a]	92.67[a]	94.67[a]	±0.67	94.00[a]	96.50[a]	94.57[a]	±0.48
Performance index	2798.20[a]	4207.58[a]	4536.36[a]	±724.51	5507.23[a]	8550.75[a]	8056.34[a]	±2082.24

[abc] means in the same row with the same superscript are not significantly ($P > 0.05$) different

SEM: Pooled Standard Error of Mean

A qualitative evaluation represented by overall scoring of performance parameters measured showed that the lowest mean score was 1.17 for D1 group in both the starter and finisher phases while the highest mean score was 2.83 and 2.67 all recorded in D3 group in the two phases of the experiment respectively.

The observed similarities in most of the performance evaluation traits in both the starter and finisher phases was as a result of the absence of major anti nutritional factors as reported by [14] and [2] and the nutrient balance of the experimental diets [17] at both the starter and finisher phases. Another factor of interest is the nutritional characteristics of Roselle seeds which supports the bioavailability of nutrients, particularly digestible protein and digestible energy [6, 7].

A qualitative comparison of the dietary treatment groups showed the nutritional superiority of fermented Roselle seed based diet. This was due to the fermentation process which is recognized as converter of food compounds into structurally related but financially more viable food through the activities of microbial cells [2]. This finding is supported by various reports on the effect of different fermentation methods on the health and growth responses of broiler chickens [1]. The synergistic effect of beneficial fermentation microbes and host microorganisms must have led to reduction in the count of pathogenic bacteria and increased the population of useful micro flora in gut, resulting to improvement in the gastrointestinal health and performance of boilers as reported by [14] and [2]. Other factors that may be responsible could be the positive effects of fermentation originating from the enzymes of the seed itself.

4. CONCLUSION

This study showed that Roselle seed is a rich source of nutrients that will provide useful replacement of conventional oilseeds in broiler feeds. Fermentation of Roselle (Hibiscus sabdarif) is thus an effective mechanism for the improvement of performance traits of broilers. Therefore, fermentation could be introduced as a safe and natural process for improving the utilization of Roselle seeds in broiler diets.

ACKNOWLEDGEMENTS

The authors sincerely acknowledgement the support of Mr Tsaku and Ramatu Aliyu of the Microbiology and Biochemistry laboratories of the Nasarawa State University Keffi, Mr Mathias Bello of the Livestock Complex of the College of Agriculture, Lafia.

COMPETING INTERESTS

Authors have declared that no competing interests exist.

REFERENCE

1.   Ari, MM and Ayanwale, BA. Nutrient Retention and Serum Profile of Broilers Fed Fermented African Locust Beans (Parkia filicoide). Asian Journal of Agricultural Research. 2012; 6(3):129-136 DOI:10.3923/ajar.2012.

2.   Ari MM, Ayanwale BA, Adama TZ, Olatunji EA. Effects of Different Fermentation Methods on the Proximate Composition, Amino Acid Profile and Some Antinutritional Factors (ANFs) In Soyabeans (*Glycine Max*) Fermentation Technology and Bioengineering. 2012a; 2 (6-13).

      Available: http://www.woaj.org/published_pdf/FTB-274.pdf

3.   Kwari ID, Igwebuike JU, Mohammed ID, Diarra SS. Growth, haematology and serum chemistry of broiler chickens fed raw or differently processed sorrel (*Hibiscus sabdariffa*) seed meal in a semi-arid environment I.J.S.N. 2011;2(1): 22-27. ISSN 2229 – 6441. Available: www.scienceandnature.org

4.   Musa- Azara SI, Ogah DM, Yakubu A, Ari MM, Hassan DI. Effects of Hibiscus calyx extracts on the blood chemistry of Broiler chickens Egypt. Poult. Sci. 2013;33(I):309-312. ISSN: 1110-5623 (Print) – 2090-0570 (On line)

5.   Tounkara F, Amadou I, Le Guo-Wei, Shi Yong-Hui. Effect of boiling on the physicochemical properties ofRoselle seeds (*Hibiscus sabdariffa* L.) cultivated in Mali Afri J. of Biotechnol. 2011;10(79)18160-18166. DOI: 10.5897/AJB11.022.

6.   Abu-Tarboush HM, Ahmed SAB, Al Kahtani HA. "Some nutritional and functional properties of Karkade (*Hibiscus sabdariffa*) seed p*roducts*". Cereal Chem. 1997;74(3):352-355. Available: http://dx.doi.org/10.1094/CCHEM.1997.74.3.352.

7.   Parkouda C, Diawara B, Ouoba LII. Technology and physicochemicalcharacteristics of Bikalga, alkali*ne fermented seeds* of *Hibiscus sabdariffa.* Afr. J. Biotechnol. 2008;7(7):916-922.

8.   Ari MM, Ayanwale BA, Adama TZ, Olatunji EA. Effect of Different Fermentation Methods on Growth Indices and Serum Profile of Broiler Chickens. Journal of Biology, Agriculture and Healthcare. 2012b;2(5):78-86. ISSN (Paper) 2224-3208 ISSN (Online) 2225-093X.

9.   Ari MM, Barde RE, Ogah MD , Yakubu A, Aya VE. Performance of Broilers Fed Silk Cotton Seed (*Ceiba Petandra*) Based Diets PAT. 2011;7(2)20-28. ISSN: 0794-5213. Available: www.patnsukjournal.net/currentissue.

10.   AOAC. Official Methods of ^Analysis, 15th Edn.Association of Official Analytical Chemists: Washington D.C; 1995.

11.   Duncan DB. Multiple rang^e and multiple F-test. Biometrics. 1955;11:1-42.

12.   SPSS. Statistical packa^ge for social science 16 0 Brief Guide: SPSS Inc. 233 South Wacker Drive, 11th Floor Chicago, IL 60606-6412 16. 2007.

13.   Asika N. Research meth^odology i*n* the Behavioural Sciences. Longman Publishers. 1991;58–68.

14.   Yagoub Abu, El Gasim A, Mohammed A. Mohammed. Furundu, a Meat Substitute from Fermented Roselle (*Hibiscus sabdariffa* L.) Seed: Investigation on Amino Acids Composition, Protein Fractions, Minerals Content and HCl-Extractability and Microbial Growth Pakistan Journal of Nutrition. 2008;7(2):352-358.

15.   Nzikou JM, Bouanga-Kalou G, Matos L, Ganongo-Po FB, Mboungou-Mboussi PS. Characteristics and Nutritional Evaluation of seed oil from Roselle (*Hibiscus sabdariffa* L.) in Congo-Brazzaville Curr. Res. J. Biol. Sci. 2011;3(2):141-146, ISSN: 2041-0778.

16.   Shaheen MA, El-Nakhlawy FS, Al-Shareef AR. Roselle (*Hibiscus sabdariffa* L.) seeds as unconventional nutritional source. Afr. J. Biotechnol. 2012;11(41):9821-9824. DOI: 10.5897/AJB11.4040.

17. NRC (National Research Council), Nutrient requirements of poultry. 9th Rev.(ed).National Academy Press, Washington, D. C; 1996.

18. Oluyemi JA, Roberts FA. Poultry production in warm wet climates Macmillan Low Cost Editions. 2000;1-145.

Permissions

All chapters in this book were first published in BBJ, by SCIENCE DOMAIN International; hereby published with permission under the Creative Commons Attribution License or equivalent. Every chapter published in this book has been scrutinized by our experts. Their significance has been extensively debated. The topics covered herein carry significant findings which will fuel the growth of the discipline. They may even be implemented as practical applications or may be referred to as a beginning point for another development.

The contributors of this book come from diverse backgrounds, making this book a truly international effort. This book will bring forth new frontiers with its revolutionizing research information and detailed analysis of the nascent developments around the world.

We would like to thank all the contributing authors for lending their expertise to make the book truly unique. They have played a crucial role in the development of this book. Without their invaluable contributions this book wouldn't have been possible. They have made vital efforts to compile up to date information on the varied aspects of this subject to make this book a valuable addition to the collection of many professionals and students.

This book was conceptualized with the vision of imparting up-to-date information and advanced data in this field. To ensure the same, a matchless editorial board was set up. Every individual on the board went through rigorous rounds of assessment to prove their worth. After which they invested a large part of their time researching and compiling the most relevant data for our readers.

The editorial board has been involved in producing this book since its inception. They have spent rigorous hours researching and exploring the diverse topics which have resulted in the successful publishing of this book. They have passed on their knowledge of decades through this book. To expedite this challenging task, the publisher supported the team at every step. A small team of assistant editors was also appointed to further simplify the editing procedure and attain best results for the readers.

Apart from the editorial board, the designing team has also invested a significant amount of their time in understanding the subject and creating the most relevant covers. They scrutinized every image to scout for the most suitable representation of the subject and create an appropriate cover for the book.

The publishing team has been an ardent support to the editorial, designing and production team. Their endless efforts to recruit the best for this project, has resulted in the accomplishment of this book. They are a veteran in the field of academics and their pool of knowledge is as vast as their experience in printing. Their expertise and guidance has proved useful at every step. Their uncompromising quality standards have made this book an exceptional effort. Their encouragement from time to time has been an inspiration for everyone.

The publisher and the editorial board hope that this book will prove to be a valuable piece of knowledge for researchers, students, practitioners and scholars across the globe.

List of Contributors

Khaled M. Aboshanab
Department of Microbiology and Immunology, Faculty of Pharmacy, Ain Shams University, Organization of African Unity St., POB: 11566, Abbassia, Cairo, Egypt

Nisreen M. Okba
Department of Microbiology and Immunology, Faculty of Pharmacy, Al-Azhar University (Girls), Cairo, Egypt

Tarek S. El-banna
Department of Pharmaceutical Microbiology, Faculty of Pharmacy, Tanta University, Tanta, Egypt

Ahmed A. Abd El-Aziz
Department of Pharmaceutical Microbiology, Faculty of Pharmacy, Tanta University, Tanta, Egypt

Gaddaguti Venu Gopal
Centre for Plant Tissue Culture and Breeding, Department of Biotechnology, K L University, Guntur- 522 502, A. P, India

Srideepthi Repalle
Centre for Plant Tissue Culture and Breeding, Department of Biotechnology, K L University, Guntur- 522 502, A. P, India

Venkateswara Rao Talluri
Centre for Plant Tissue Culture and Breeding, Department of Biotechnology, K L University, Guntur- 522 502, A. P, India

Srinivasa Reddy Ronda
Centre for bioprocess Technology, Department of Biotechnology, K L University, Guntur- 522 502, A.P, India

Prasada Rao Allu
Centre for Plant Tissue Culture and Breeding, Department of Biotechnology, K L University, Guntur- 522 502, A. P, India

P. V. Krishna
Department of Zoology and Aquaculture, Acharya Nagarjuna University, Nagarjuna Nagar-522 510, Andhra Prades, India

K. Madhusudhana Rao
Department of Zoology and Aquaculture, Acharya Nagarjuna University, Nagarjuna Nagar-22 510, Andhra Prades, India

V. Swaruparani
Department of Zoology and Aquaculture, Acharya Nagarjuna University, Nagarjuna Nagar-522 510, Andhra Prades, India

D. Srinivas Rao
Departmrnt of Biotechnology, Acharya Nagarjuna University, Nagarjuna Nagar-522 510, Andhra Prades, India

O. C Eruteya
Microbiology Department, University of Port Harcourt, Port Harcourt, Nigeria

S. A Odunfa
Department of Microbiology, University of Ibadan, Ibadan, Nigeria

J. Lahor
Microbiology Department, University of Port Harcourt, Port Harcourt, Nigeria

L. F. C. Ribeiro
Immunology and Biochemistry Department of Faculdade de Medicina de Ribeirão Preto – USP, Ribeirão Preto, SP, Brazil

L. F. Ribeiro
Immunology and Biochemistry Department of Faculdade de Medicina de Ribeirão Preto – USP, Ribeirão Preto, SP, Brazil

J. A. Jorge
Biology Department of Faculdade de Filosofia Ciências e Letras de Ribeirão Preto – USP, Ribeirão Preto, SP, Brazil

M. L. T. M. Polizeli
Biology Department of Faculdade de Filosofia Ciências e Letras de Ribeirão Preto – USP, Ribeirão Preto, SP, Brazil

L. O. Okwute
Department of Biological Sciences, University of Abuja, P. M. B. 117, Gwagwalada-Abuja, Nigeria

U. J. J. Ijah
Department of Microbiology, Federal University of Technology, P. M. B. 65, Minna-Niger State, Nigeria

Charles Ogugua Nwuche
Department of Microbiology, Faculty of Biological Sciences, University of Nigeria, Nsukka 40001, Nigeria
Department of Bioscience and Bioengineering, Graduate School of Life and Environmental Sciences, University of Tsukuba, Ibaraki, Japan

Doris Chidimma Ekpo
Department of Microbiology, Faculty of Biological Sciences, University of Nigeria, Nsukka 40001, Nigeria

Chijioke Nwoye Eze
Department of Microbiology, Faculty of Biological Sciences, University of Nigeria, Nsukka 40001, Nigeria

Hideki Aoyagi
Department of Bioscience and Bioengineering, Graduate School of Life and Environmental Sciences, University of Tsukuba, Ibaraki, Japan

James Chukwuma Ogbonna
Department of Microbiology, Faculty of Biological Sciences, University of Nigeria, Nsukka 40001, Nigeria
Department of Bioscience and Bioengineering, Graduate School of Life and Environmental Sciences, University of Tsukuba, Ibaraki, Japan

Farrokhi Meherdad
Environmental Health Department, School of health, Guilan University of Medical Sciences, Rasht, Iran

Ghaemi Naser
Biochemistry & Biotechnology Department, Tehran University of Medical Sciences, Tehran, Iran

Najafi Fazel
School of base of science, Guilan University, Rasht, Iran

Naimi- Joubani Mohammad
Environmental Health Department, School of health, Guilan University of Medical Sciences, Rasht, Iran

Farmanbar Rabiollah
Health Education Department, School of health, Guilan University of Medical Sciences, Rasht, Iran

Roohbakhsh Joorshari Esmaeil
Health Education Department, School of health, Guilan University of Medical Sciences, Rasht, Iran

T. Santra
Department of Microbiology, Institute of Genetic Engineering, Badu, Kol-128, India

S. K. Ghosh
Molecular Mycopathology Lab, P. G. Department of Botany, Ramakrishna Mission Vivekananda Centenary College, Rahara, Kol-118, India

A. Chakravarty
Department of Molecular biology, Institute of Genetic Engineering, Badu, Kol-128, India

O. A. Oseni
Department of Medical Biochemistry, College of Medicine, Ekiti State University, Ado-Ekiti, Nigeria

M. M. Ekperigin
Department of Biochemistry, Federal University of Technology, Akure, Ondo State, Nigeria

Btissam Ben Messaoud
Soil and Environment Microbiology Unit, Faculty of Sciences, Moulay Ismail University, Meknes, Morocco

Imane Aboumerieme
Soil and Environment Microbiology Unit, Faculty of Sciences, Moulay Ismail University, Meknes, Morocco

Laila Nassiri
Soil and Environment Microbiology Unit, Faculty of Sciences, Moulay Ismail University, Meknes, Morocco

Elmostafa El Fahime
Technical Support Unit, Scientific Research, CNRST in Rabat, Morocco

Jamal Ibijbijen
Soil and Environment Microbiology Unit, Faculty of Sciences, Moulay Ismail University, Meknes, Morocco

Maikano Mohammed Ari
Department of Animal Science, Faculty of Agriculture, Nasarawa State University Keffi, Shabu- Lafia Campus, Nasarawa State, Nigeria

Danlami Moses Ogah
Department of Animal Science, Faculty of Agriculture, Nasarawa State University Keffi, Shabu- Lafia Campus, Nasarawa State, Nigeria

Idris Danladi Hassan
Department of Animal Production, College of Agriculture Lafia, Nasarawa State, Nigeria

Ibrahim Suleiman Musa-Azara
Department of Animal Production, College of Agriculture Lafia, Nasarawa State, Nigeria

Nuhu Dalami Yusuf
Department of Animal Production, College of Agriculture Lafia, Nasarawa State, Nigeria

Samuel Emmanuel Alu
Department of Animal Science, Faculty of Agriculture, Nasarawa State University Keffi, Shabu- Lafia Campus, Nasarawa State, Nigeria